IMAGE PROCESSING FOR CINEMA

CHAPMAN & HALL/CRC MATHEMATICAL AND COMPUTATIONAL IMAGING SCIENCES

Series Editors

Chandrajit Bajaj
Center for Computational Visualization
The University of Texas at Austin

Guillermo Sapiro
Department of Electrical
and Computer Engineering
Duke University

Aims and Scope

This series aims to capture new developments and summarize what is known over the whole spectrum of mathematical and computational imaging sciences. It seeks to encourage the integration of mathematical, statistical and computational methods in image acquisition and processing by publishing a broad range of textbooks, reference works and handbooks. The titles included in the series are meant to appeal to students, researchers and professionals in the mathematical, statistical and computational sciences, application areas, as well as interdisciplinary researchers involved in the field. The inclusion of concrete examples and applications, and programming code and examples, is highly encouraged.

Published Titles

Image Processing for Cinema
by Marcelo Bertalmío

Statistical and Computational Methods in Brain Image Analysis
by Moo K. Chung

Rough Fuzzy Image Analysis: Foundations and Methodologies
by Sankar K. Pal and James F. Peters

Theoretical Foundations of Digital Imaging Using MATLAB®
by Leonid P. Yaroslavsky

Proposals for the series should be submitted to the series editors above or directly to:
CRC Press, Taylor & Francis Group
3 Park Square, Milton Park, Abingdon, OX14 4RN, UK

CHAPMAN & HALL/CRC
MATHEMATICAL AND COMPUTATIONAL IMAGING SCIENCES

IMAGE PROCESSING FOR CINEMA

MARCELO BERTALMÍO
Universitat Pompeu Fabra
Barcelona, Spain

CRC Press is an imprint of the
Taylor & Francis Group, an **informa** business
A CHAPMAN & HALL BOOK

CRC Press
Taylor & Francis Group
6000 Broken Sound Parkway NW, Suite 300
Boca Raton, FL 33487-2742

First issued in paperback 2019

ISBN-13: 978-1-4398-9927-4 (hbk)
ISBN-13: 978-0-367-37893-6 (pbk)

Visit the Taylor & Francis Web site at
http://www.taylorandfrancis.com

and the CRC Press Web site at
http://www.crcpress.com

Para Lucas y Vera,
y Guillermo y Gregory,
y papá y Serrana.
Siempre presente, Vicent.

Acknowledgments

The original illustrations in this book are by killer friends and knock-out artists Javier Baliosian, Federico Lecumberry and Jorge Visca. They also helped me with the book cover, along with Rafael Grompone. The rest of the figures are reproduced with the permission of their authors, to whom I'm very grateful.

A heartfelt "Thank you" to all the researchers I've had the pleasure of collaborating with over the years: Gregory Randall, Guillermo Sapiro, Coloma Ballester, Stan Osher, Li Tien Cheng, Andrea Bertozzi, Alicia Fernández, Shantanu Rane, Luminita Vese, Joan Verdera, Oliver Sander, Pere Fort, Daniel Sánchez-Crespo, Kedar Patwardhan, Juan Cardelino, Gloria Haro, Edoardo Provenzi, Alessandro Rizzi, Álvaro Pardo, Luis Garrido, Adrián Marques, Aurélie Bugeau, Sira Ferradans, Rodrigo Palma-Amestoy, Jack Cowan, Stacey Levine, Thomas Batard, Javier Vazquez-Corral, David Kane, Syed Waqas Zamir and Praveen Cyriac.

Special thanks to: Sunil Nair, Sarah Gelson, Michele Dimont and everyone at Taylor & Francis, and Aurelio Ruiz.

Very special thanks to Jay Cassidy, Pierre Jasmin and Stan Osher.

Finally, I want to acknowledge the support of the European Research Council through the Starting Grant ref. 306337, of ICREA through their ICREA Acadèmia Award, and of the Spanish government through the grants TIN2012-38112 and TIN2011-1594-E.

Always present: Andrés Solé and Vicent Caselles.

Preface

This book is intended primarily for advanced undergraduate and graduate students in applied mathematics, image processing, computer science and related fields, as well as for researchers from academia and professionals from the movie industry. It can be used as a textbook for a graduate course, for an advanced undergraduate course, or for summer school. My intention has been that it can serve as a self-contained handbook and a detailed overview of the relevant image processing techniques that are used in practice in cinema. It covers a wide range of topics showing how image processing has become ubiquitous in movie-making, from shooting to exhibition. It does not deal with visual effects or computer-generated images, but rather with all the ways in which image processing algorithms are used to enhance, restore, adapt or convert moving images, their purpose being to make the images look as good as possible while exploiting all the capabilities of cameras, projectors and displays.

Image processing is by definition an applied discipline, but very few of the image processing algorithms intended for application in the cinema industry are ever actually used. There probably are many reasons for this, but in my view the most important ones are that we, the researchers, are often not aware of the impossibly high quality standards of cinema, and also we don't have a clear picture of what the needs of the industry are, what problems they'd really like to solve versus what we *think* they'd like to solve.

Movie professionals, on the other hand, are very much aware of what their needs are, they are very eager to learn and try new techniques that may help them and they want to *understand* what is it they are applying. But very often the technical or scientific information they want is spread over many texts, or buried under many layers of math or unrelated exposition.

Surprisingly, then, this is the first comprehensive book on image processing for cinema, and I've written it because I sincerely think that having all this information together can be beneficial both for researchers and for movie industry professionals.

Current digital cinema cameras match or even surpass film cameras in color capabilities, dynamic range and resolution, and several of the largest camera makers have ceased production of film cameras. On the exhibition side, film is forecasted to be gone from American movie theaters by 2015. And while many mainstream and blockbuster movies are still being shot on film, they are all digitized for postproduction. For all these reasons this book equates "cinema" with "digital cinema," considers only digital cameras and digital movies, and does not deal with image processing algorithms for problems that are inherent to film, like the restoration of film scratches or color fading.

The book is structured in three parts. The first one covers some fundamentals on optics and color. The second part explains how cameras work and details all the image processing algorithms that are applied in-camera. The last part is devoted to image processing algorithms that are applied off-line in order to solve a wide range of problems, presenting state-of-the-art methods with special emphasis on the techniques that are actually used in practice. The mathematical presentation of all methods will concentrate on their purpose and idea, leaving formal proofs and derivations for the interested reader in the cited references.

Finally: I've written this book in the way I like to read (technical) books. I hope you enjoy it.

Marcelo Bertalmío
Barcelona, July 2013

Contents

Part I

Lights

Chapter 1

Light and color

1.1 Light as color stimulus

We live immersed in electromagnetic fields, surrounded by radiation of natural origin or produced by artifacts made by humans. This radiation has a dual behavior of wave and particle, where the particles can be considered as packets of electromagnetic waves. Waves are characterized by their wavelength, the distance between two consecutive peaks. Of all the radiation that continuously reaches our eyes, we are only able to *see* (i.e. our retina photoreceptors are only sensitive to) electromagnetic radiation with wavelengths within the range of 380nm to 740nm (a nm or nanometer is one-billionth of a meter). We are not able to see radiation outside this band, such us ultraviolet radiation (wavelength of 10nm to 400nm) or FM radio (wavelengths near 1m). Therefore, light is defined as radiation with wavelengths within the *visible spectrum* of 380nm to 740nm. Figure 1.1 shows the full spectrum of radiation with a detail of the visible light spectrum. The sun emits full-spectrum radiation, including gamma rays and ultraviolet and infrared "light," which of course have an effect on our bodies even if we are not able to see them.

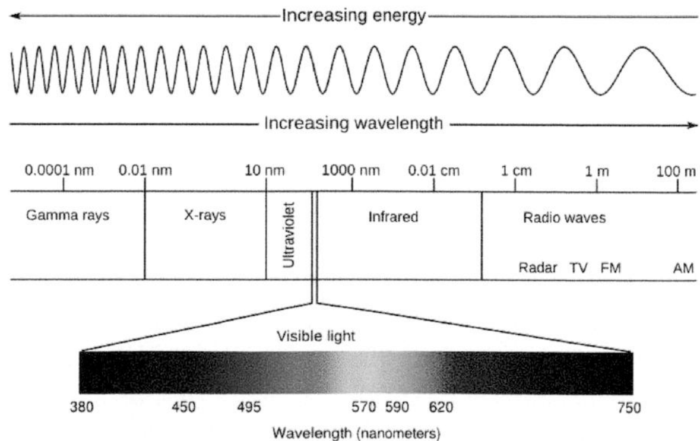

FIGURE 1.1: Electromagnetic spectrum and visible light.

The frequency of a wave is its number of cycles per second; all electromagnetic waves travel at the same speed, the speed of light (c in the vacuum), therefore the wavelength of a wave is inversely proportional to its frequency. Shorter wavelengths imply higher frequencies and also higher energies. When we look at a *single isolated light*, if it has short wavelength we perceive it as blue, if it has middle-length wavelength we see it as green, and if it has long wavelength it appears to us as red. But we must stress that light in itself is not colored (there are no different kinds of photons), color is a perceptual quantity: for instance, the same light that appears red when isolated may appear yellow when it is surrounded by other lights. So the light stimulus at a given location in the retina is not enough to determine the color appearance it will produce; nonetheless, it must be characterized since it constitutes the input to our visual system and therefore what color appearance will depend on. Among other ways, light stimuli can be described by radiometry, which measures light in energy units and does not consider the properties of our visual system, and by colorimetry, which reduces the multi-valued radiometric spectrum of a light stimulus to three values describing the effect of the stimulus in the three types of cone receptors in the retina [298].

With a radiometric approach, the properties of a light emitting source are described by its power spectrum function $I(\lambda)$, the *irradiance*, which states for each wavelength λ the amount of power I the light has. The light absorption properties of a surface are described by its *reflectance* $R(\lambda)$, which for each wavelength λ states the percentage of photons that are reflected by the surface. When we see a surface, the light that is reflected by it and reaches our eyes is called *radiance* and its power spectrum $E(\lambda)$ is the product of the spectrum functions for the incident light and the reflectance function of the surface:

$$E(\lambda) = I(\lambda) \times R(\lambda). \tag{1.1}$$

Figure 1.2 shows the irradiance functions of several light emitting sources, and Figure 1.3 shows the reflectances of some patches of different colors. From these figures and Equation 1.1 we can see that when we illuminate a red patch (Figure 1.3(a)) with sunlight (Figure 1.2(a)) we get from the patch a radiance E with its power concentrated in the longest wavelengths, which as we mentioned corresponds to our sensation of red.

The human retina has photoreceptor neurons, with colored pigments. These pigments have their particular photon absorption properties as a function of wavelength, and absorbed photons generate chemical reactions that produce electrical impulses, which are then processed at the retina and later at the visual cortex in the brain. The sensitivity of the pigments depends on the *luminance* of the light, which is a measure of the light's power, formally defined as intensity per unit area in a given direction. We have two types of phororeceptors:

– Rods, for low and mid-low luminances (at high luminances they are active but saturated). There are some 120 million of them.

– Cones, which have pigments that are 500 times *less* sensitive to light

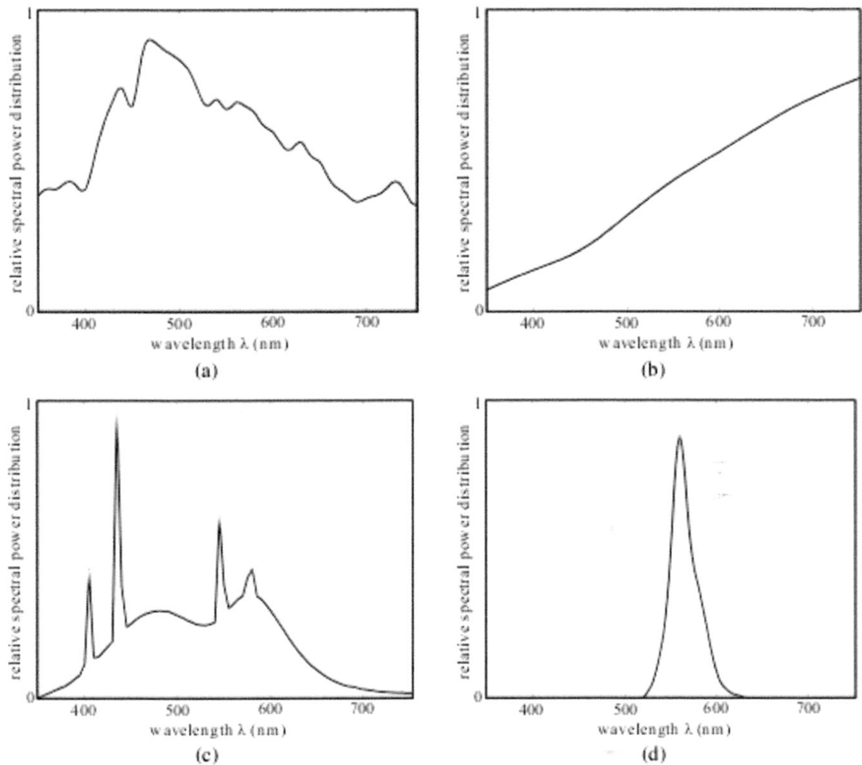

FIGURE 1.2: Spectral power distribution of various common types of illuminations: (a) sunlight, (b) tungsten light, (c) fluorescent light, and (d) LED. Figure from [239].

than the rods' pigment, rhodopsin. Therefore, cones work only with high luminances; at low luminances they are not active. There are some 6 million of them, most of them very densely concentrated at the fovea, the center of the retina.

There are three types of cones: S-cones, M-cones and L-cones, where the capital letters stand for "short," "medium" and "long" wavelengths, respectively. Hubel [203] points out that three is the minimum number of types of cones that allow us not to confuse any monochromatic light with white light. People who lack one type of cone do perceive certain colors as gray. Frisby and Stone [178] mention that while there are several animal species with more than three types of color receptors, which can then tell apart different shades of color that we humans perceive as equal, this probably comes at the price of less visual acuity, for there are more cones to be accomodated in the same retinal area. Low luminance or *scotopic* vision, mediated only by rods, is there-

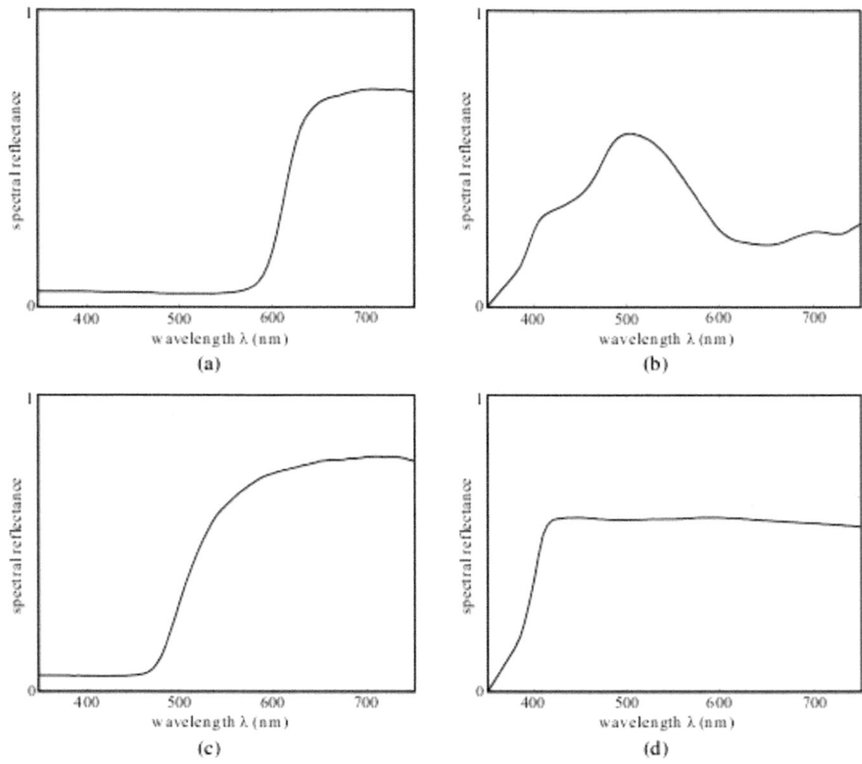

FIGURE 1.3: Spectral reflectance of various colored patches: (a) red patch, (b) blue patch, (c) yellow patch, and (d) gray patch. Figure from [239].

fore color-less. In a low-medium range of luminances, the so-called *mesopic* vision, both rods and cones are active, and this is what happens in a typical movie theatre [86]. In high-luminance or *photopic* vision cones are active and the rods are saturated. Each sort of cone photoreceptor has a spectral absorbance function describing its sensitivity to light as a function of wavelength: $s(\lambda), m(\lambda), l(\lambda)$. These curves were first determined experimentally by König in the late 19th century. The sensitivity curves are quite broad, almost extending over the whole visible spectrum, but they are bell-shaped and they peak at distinct wavelengths: S-cones at 420nm, M-cones at 533nm and L-cones at 584nm; see Figure 1.4. These three *wavelength* values correspond to monochromatic blue, green and red light, respectively.

With the colorimetric approach, the sensation produced in the eye by radiance $E(\lambda)$ (the stimulus of a light of power spectrum $E(\lambda)$) is determined by a triplet of values, called the tristimulus values, given by the integral over

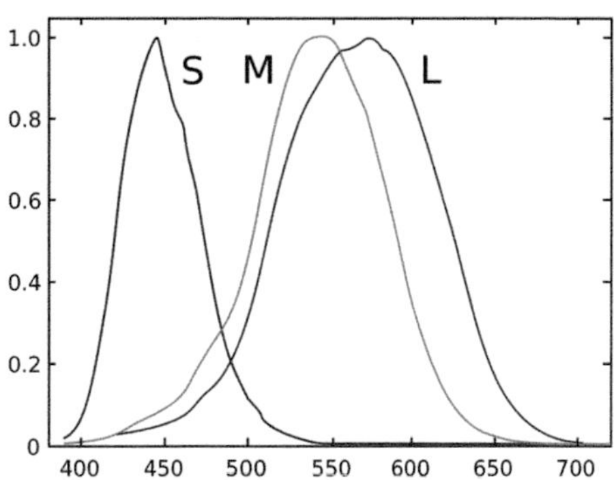

FIGURE 1.4: Cone sensitivities (normalized). Figure from [19].

the visible spectrum of the product of the radiance by each of the three cone sensitivity functions:

$$
\begin{aligned}
L &= \int_{380}^{740} l(\lambda)E(\lambda)d\lambda \\
M &= \int_{380}^{740} m(\lambda)E(\lambda)d\lambda \\
S &= \int_{380}^{740} s(\lambda)E(\lambda)d\lambda.
\end{aligned} \tag{1.2}
$$

1.2 Matching colors

If two lights with different spectra $E_1(\lambda)$ and $E_2(\lambda)$ produce the same tristimulus vector (L, M, S), then both lights are producing the same cone responses and (if viewed in isolation) we will see them as having the same color. Lights with different spectra that appear to have the same color are called metamers (lights with the same spectra always appear to have the same color and are called isomers [214]).

Expressing Equation 1.2 in discrete terms, it will be easy to see that each light has many metamers:

$$L = \sum_{i=380}^{740} l(\lambda_i)E(\lambda_i)$$
$$M = \sum_{i=380}^{740} m(\lambda_i)E(\lambda_i)$$
$$S = \sum_{i=380}^{740} s(\lambda_i)E(\lambda_i), \tag{1.3}$$

which, arranged into matrix form, becomes:

$$\begin{bmatrix} L \\ M \\ S \end{bmatrix} = \mathcal{S}\mathbf{E}, \tag{1.4}$$

where $\mathcal{S}$ is a 3×361 matrix whose rows are the (discrete) cone sensitivities, and $\mathbf{E}$ is the radiance spectrum expressed as a column vector. Equation 1.4 is a (very) undetermined system of equations, with only three equations for the 361 unknowns of the radiance vector, and therefore for every light $\mathbf{E_1}$ there will be many lights $\mathbf{E_2}$ producing the same tristimulus (L, M, S), i.e. many metamers.

From Equation 1.4 we can also derive the property of *trichromacy*, which is a fundamental property of human color vision: simply put, it means that we can generate any color by mixing three given colors, merely adjusting the amount of each. In his excellent account of the origins of color science, Mollon [276] explains how this was a known fact and how it was already applied in the 18th century, for printing in full color using only three types of colored ink. Trichromacy is due to our having three types of cone photoreceptors in the retina, therefore we must remark that it is *not* a property of light but a property of our visual system. This was not known in the 18th century: light sensation was supposed to be transmitted directly along the nerves, so trichromacy was thought to be a physical characteristic of the light. Thomas Young was the first to explain, in 1801, that the variable associated with color in light is the wavelength and, since it varies continously, the trichromacy must be imposed by the visual system and hence there must be three kind of receptors in the eye. He was also the first to realize that visible light is simply radiation within a certain waveband, and radiation with freqencies outside this range was not visible but could be felt as heat.

Following the approach in [341], we can take any three primaries of spectra $\mathbf{P_i}$,i=1,2,3, as long as they are colorimetrically independent, meaning that none of the three lights can be matched in color by a combination of the other two lights. Let $\mathbf{P}$ be a $N \times 3$ matrix whose columns are $\mathbf{P_i}$. Given a stimulus light of power spectrum $\mathbf{E}$, we compute the following three-element vector $\mathbf{w}$:

$$\begin{bmatrix} w_1 \\ w_2 \\ w_3 \end{bmatrix} = \mathbf{w} = (\mathcal{S}\mathbf{P})^{-1}\mathcal{S}\mathbf{E}. \tag{1.5}$$

Pre-multiplying each side of the equation by $\mathcal{S}\mathbf{P}$, we get:

$$\mathcal{S}\mathbf{P}\mathbf{w} = \mathcal{S}\mathbf{E}, \tag{1.6}$$

which, according to Equation 1.4, means that lights **Pw** and **E** are metamers. But light **Pw** is just a linear combination of the primaries $\mathbf{P_i}$ with weights w_i:

$$\mathbf{P}\mathbf{w} = (\mathbf{w_1}\mathbf{P_1} + \mathbf{w_2}\mathbf{P_2} + \mathbf{w_3}\mathbf{P_3}). \tag{1.7}$$

Therefore, any light stimulus **E** can be matched by a mixture of any three (colorimetrically independent) lights $\mathbf{P_i}$ with the intensities adjusted by w_i, which is the property of trichromacy.

But, for any set **P** of primaries, there are always some lights **E** for which the weight vector **w** has a negative component [341]. While mathematically this is not a problem, physically it makes no sense to have lights of negative intensity, so in these cases it is *not* possible to match **E** with the primaries. What is physically realizable, though, is to match the mixture of **E** and the primary of the negative weight with the other two primaries. For instance, if the weight vector is $w_1 = -\alpha, w_2 = \beta, w_3 = \gamma$, with $\alpha, \beta, \gamma > 0$, i.e. the first primary has negative weight, then from equations 1.6 and 1.7 we get:

$$\mathcal{S}(\beta\mathbf{P_2} + \gamma\mathbf{P_3}) = \mathcal{S}(\alpha\mathbf{P_1} + \mathbf{E}). \tag{1.8}$$

In this example, then, light **E** cannot be matched by any mixture of $\mathbf{P_1}, \mathbf{P_2}, \mathbf{P_3}$, but if we add $\alpha\mathbf{P_1}$ to **E** we can match this light to $\beta\mathbf{P_2}+\gamma\mathbf{P_3}$. This is precisely what Maxwell observed in his seminal color matching experiments, in 1855-1860, which served as the basis for all technological applications involving color acquisition and reproduction, from cameras to displays to printing. He presented an observer with a monochromatic test light of wavelength λ, and the observer had to match its color by varying the intensity of red, green and blue lights. In the cases when this wasn't posible, the red or the blue light were added to the test light so that the other two primaries were matched to the mixture of the original test light and the remaining primary. For each λ Maxwell recorded the weights w_1, w_2, w_3, which produced the color match. The weights are also a function of λ: the functions $w_i(\lambda),\ i = 1, 2, 3$, are called the color matching functions for the primaries. In 1861, in a lecture before the Royal Society of London, Maxwell created the first color photograph: he took three black and white photographs of the same object, in each occasion placing a different colored filter (red, green or blue) before the camera; then, each photograph was projected onto the same screen, using three projectors with the corresponding colored filters before them. The final image observed on the screen reproduced, although imperfectly, the colors of the photographed object [86]. In the late 1920's, W. David Wright (and, independently, John Guild) conducted the same experiment with a group of observers, asking them to color-match a given monochromatic light by varying the brightness of a set of red, green and blue monochromatic lights [384]. For each monochromatic test light of wavelength λ, the experiment recorded the (average, over all observers) amounts of each primary needed to match the test: $\bar{r}(\lambda)$ for the red, $\bar{g}(\lambda)$ for the green, and $\bar{b}(\lambda)$ for the blue. These are the color matching functions for a standard observer.

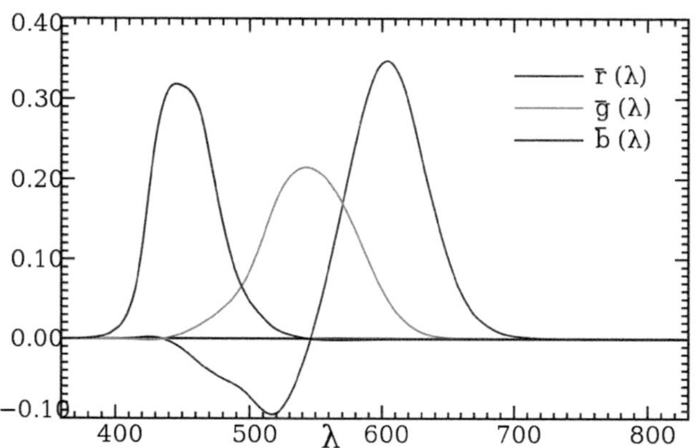

FIGURE 1.5: Color matching functions. Figure from [13].

We can see in Figure 1.5 that the $\bar{r}(\lambda)$ function clearly has some negative values (the other two functions have negative values as well). As we explained above, these correspond to cases when the test color can't be matched as it is, unless a certain amount of primary color (red) is added to it, in which case the brightness for the green and blue lights can be adjusted so as to match the modified test light.

From the color matching functions, the (R, G, B) tristimulus values for a light source with spectral distribution $E(\lambda)$ can be computed as:

$$\begin{aligned} R &= \int_{380}^{740} \bar{r}(\lambda)E(\lambda)d\lambda \\ G &= \int_{380}^{740} \bar{g}(\lambda)E(\lambda)d\lambda \\ B &= \int_{380}^{740} \bar{b}(\lambda)E(\lambda)d\lambda, \end{aligned} \tag{1.9}$$

or, in discrete terms:

$$R = \sum_{i=380}^{740} \bar{r}(\lambda_i)E(\lambda_i)$$

$$G = \sum_{i=380}^{740} \bar{g}(\lambda_i)E(\lambda_i)$$

$$B = \sum_{i=380}^{740} \bar{b}(\lambda_i)E(\lambda_i). \tag{1.10}$$

It is easy to show [341] that, since every light stimulus can be expressed as a mixture of monochromatic stimuli $E(\lambda_i)$ and these in turn can be color-matched to a linear combination of primaries with weights given by $\bar{r}(\lambda_i), \bar{g}(\lambda_i)$ and $\bar{b}(\lambda_i)$, then it follows that the color matching functions are a linear combination of the cone sensitivity functions $l(\lambda), m(\lambda), s(\lambda)$. We will now prove this statement, following [341]; let $\mathbf{e_i}$ be a monochromatic light of wavelength λ_i, i.e. the vector $\mathbf{e_i}$ will be 1 at position i and 0 elsewhere. Then, from Equation 1.4:

$$\begin{bmatrix} L_i \\ M_i \\ S_i \end{bmatrix} = \mathcal{S}\mathbf{e_i} = \begin{bmatrix} l(\lambda_i) \\ m(\lambda_i) \\ s(\lambda_i) \end{bmatrix}, \tag{1.11}$$

and from Equation 1.6:

$$\begin{aligned} \mathcal{S}\mathbf{e_i} = \mathcal{S}\mathbf{P}\mathbf{w} = \mathcal{S}\left(\bar{\mathbf{r}}(\lambda_\mathbf{i})\mathbf{R} + \bar{\mathbf{g}}(\lambda_\mathbf{i})\mathbf{G} + \bar{\mathbf{b}}(\lambda_\mathbf{i})\mathbf{B}\right) = \\ \bar{r}(\lambda_i)\mathcal{S}\mathbf{R} + \bar{\mathbf{g}}(\lambda_\mathbf{i})\mathcal{S}\mathbf{G} + \bar{\mathbf{b}}(\lambda_\mathbf{i})\mathcal{S}\mathbf{B} = \\ \bar{r}(\lambda_i)\begin{bmatrix} L_R \\ M_R \\ S_R \end{bmatrix} + \bar{g}(\lambda_i)\begin{bmatrix} L_G \\ M_G \\ S_G \end{bmatrix} + \bar{b}(\lambda_i)\begin{bmatrix} L_B \\ M_B \\ S_B \end{bmatrix}. \end{aligned} \tag{1.12}$$

From Equations 1.11 and 1.12:

$$\begin{bmatrix} l(\lambda_i) \\ m(\lambda_i) \\ s(\lambda_i) \end{bmatrix} = \bar{r}(\lambda_i)\begin{bmatrix} L_R \\ M_R \\ S_R \end{bmatrix} + \bar{g}(\lambda_i)\begin{bmatrix} L_G \\ M_G \\ S_G \end{bmatrix} + \bar{b}(\lambda_i)\begin{bmatrix} L_B \\ M_B \\ S_B \end{bmatrix}. \tag{1.13}$$

By definition of $\mathcal{S}$:

$$\mathcal{S} = \begin{bmatrix} l(\lambda_{380}) & l(\lambda_{381}) & ... & l(\lambda_{740}) \\ m(\lambda_{380}) & m(\lambda_{381}) & ... & m(\lambda_{740}) \\ s(\lambda_{380}) & s(\lambda_{381}) & ... & s(\lambda_{740}) \end{bmatrix} = \mathcal{M}\begin{bmatrix} \bar{r}(\lambda_{380}) & \bar{r}(\lambda_{381}) & ... & \bar{r}(\lambda_{740}) \\ \bar{g}(\lambda_{380}) & \bar{g}(\lambda_{381}) & ... & \bar{g}(\lambda_{740}) \\ \bar{b}(\lambda_{380}) & \bar{b}(\lambda_{381}) & ... & \bar{b}(\lambda_{740}) \end{bmatrix}, \tag{1.14}$$

where

$$\mathcal{M} = \begin{bmatrix} L_R & L_G & L_B \\ M_R & M_G & M_B \\ S_R & S_G & S_B \end{bmatrix}. \tag{1.15}$$

Therefore, if lights $E_1(\lambda)$ and $E_2(\lambda)$ are metamers, then not only do they produce the same (L, M, S) tristimulus (as per definition of metamerism), they also produce the same (R, G, B) tristimulus. Furthermore, we can convert (R, G, B) to (L, M, S) by simple multiplication by a $3x3$ matrix $\mathcal{M}$ whose i-th column (i=1,2,3) is the (L, M, S) tristimulus value of the i-th primary, (e.g. the first column of $\mathcal{M}$ is the (L, M, S) tristimulus of the red primary), and we can convert (L, M, S) into (R, G, B) by multiplication by the inverse of $\mathcal{M}$. The matrix $\mathcal{M}$ is invertible because it is not singular, since the primaries must be colorimetrically independent, as pointed out before.

1.3 The first standard color spaces

From what has been said in the previous section, a color sensation can be described with three parameters; given a test color, we call its *tristimulus* values the amounts of three colors (primaries in some additive color model) that are needed to match that test color. If two single, isolated colored lights have different spectral distributions but the same tristimulus values, then they will be perceived as being of the same color. A *color space* is a method that associates colors with tristimulus values. Therefore, it is described by three primaries and their corresponding color matching functions, as seen in Equation 1.14.

In 1931, the International Commission on Illumination (or CIE, for its French name) amalgamated Wright and Guild's data [158] and proposed two sets of color matching functions for a standard observer, known as CIE RGB and CIE XYZ; this standard for colorimetry is still today one of the most used methods for specifying colors in the industry. The CIE RGB color matching functions are the functions $\bar{r}(\lambda), \bar{g}(\lambda), \bar{b}(\lambda)$ mentioned earlier. The tristimulus values (R, G, B) for a light $E(\lambda)$ are computed from these functions as stated in Equation 1.9.

For each wavelength λ, one of the three functions is negative. This posed a problem, since the calculators of the time were manually operated and hence errors were quite common in the computation of the tristimulus values [341]. That's why the CIE XYZ color matching functions $\bar{x}(\lambda), \bar{y}(\lambda), \bar{z}(\lambda)$ were also introduced alongside the CIE RGB ones. From the CIE XYZ functions, the (X, Y, Z) tristimulus values for a light source with spectral distribution $E(\lambda)$ can be computed as:

$$X = \int_{380}^{740} \bar{x}(\lambda)E(\lambda)d\lambda$$
$$Y = \int_{380}^{740} \bar{y}(\lambda)E(\lambda)d\lambda$$
$$Z = \int_{380}^{740} \bar{z}(\lambda)E(\lambda)d\lambda. \qquad (1.16)$$

The color matching functions $\bar{x}(\lambda), \bar{y}(\lambda), \bar{z}(\lambda)$ are obtained as a linear combination of $\bar{r}(\lambda), \bar{g}(\lambda), \bar{b}(\lambda)$ by imposing certain criteria, chiefly among them:

- $\bar{x}(\lambda), \bar{y}(\lambda), \bar{z}(\lambda)$ must always be positive;
- $\bar{y}(\lambda)$ is identical to the standard luminosity function $V(\lambda)$, which is a dimension-less function describing the sensitivity to light as a function of wavelength; therefore, $Y = \int \bar{y}(\lambda)E(\lambda)d\lambda$ would correspond to the luminance or perceived brightness of the color stimulus;
- $\bar{x}(\lambda), \bar{y}(\lambda), \bar{z}(\lambda)$ are normalized so that they produce equal tristimulus values $X = Y = Z$ for a white light, i.e. a light with a uniform (flat) spectrum.

An important point we must stress is the following. In section 1.2 we saw that for any set of physically realizable primaries there were wavelengths λ for which the color matching values were negative. Since $\bar{x}(\lambda), \bar{y}(\lambda), \bar{z}(\lambda)$ are always positive, this implies that their primaries can never be physically realizable. This is why the primaries for CIE XYZ are called *virtual primaries*.

1.3.1 Chromaticity diagrams

There are three quantities that describe our perception of a given color:

- its hue, what we normally refer to as "color" (yellow, red, and so on) and which depends on the wavelength values present in the light;
- its saturation, which refers to how "pure" it is as opposed to "how mixed with white" it is: for instance, the color red (as in blood-red) is more saturated than the color pink (red mixed with white); saturation depends on the spread of the light spectrum around its wavelength(s), a more concentrated spectrum corresponds to a more saturated color;
- its brightness, which expresses the intensity with which we perceive the color; it corresponds to the average of the power of the absorbed light.

It is usual to decouple brightness from the other quantities describing a color, and the pair hue-saturation is referred to as *chromaticity*: e.g. light blue and dark blue have the same chromaticity but different brightness.

Given their relationship to tristimulus values, color spaces are three-dimensional: each color can be represented as a point in a three-dimensional plot. See Figure 1.6 for a 3D representation of the XYZ color space.

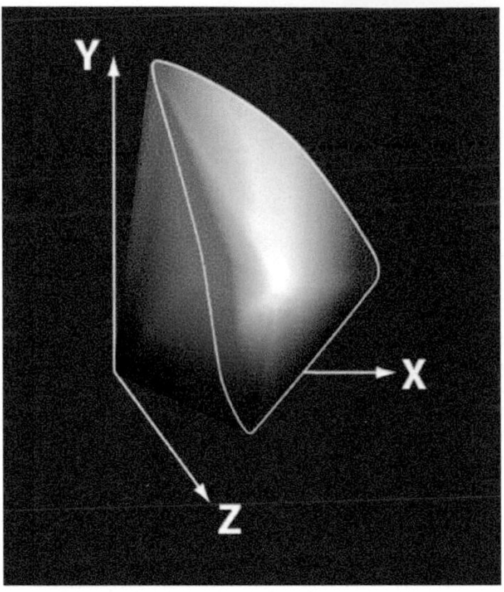

FIGURE 1.6: CIEXYZ color space. Figure from [24].

But, as we just mentioned, when describing colors it is usual to decouple luminance from chromaticity. A light stimulus $E_1(\lambda)$ and a scaled version of it $E_2(\lambda) = kE_1(\lambda)$ will produce tristimulus (X_1, Y_1, Z_1) and (kX_1, kY_1, kZ_1), respectively, which have the same chromaticity but different intensity. We now define the values x, y, z:

$$\begin{aligned} x &= \frac{X}{X+Y+Z} \\ y &= \frac{Y}{X+Y+Z} \\ z &= \frac{Z}{X+Y+Z}. \end{aligned} \tag{1.17}$$

It is easy to see that, for the abovementioned example of lights E_1 and $E_2 = kE_1$, these values are identical: $x_1 = x_2$, $y_1 = y_2$, $z_1 = z_2$. This is why x, y, z are called the *chromaticity coordinates*, because they don't change if the light stimulus only varies its intensity.

By construction $x + y + z = 1$, so $z = 1 - x - y$ and all the information of the chromaticity coordinates is contained in the pair (x, y). Therefore, all the possible chromaticities can be represented in a 2D plane, the plane with axes x and y, and this is called the CIE xy chromaticity diagram; see Figure 1.7.

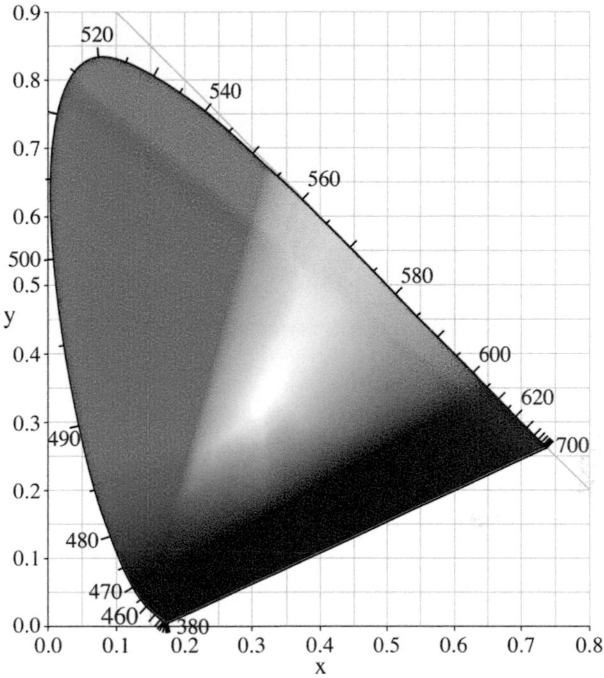

FIGURE 1.7: CIE xy chromaticity diagram. Figure from [13].

This tongue-shaped region represents all the differents chromaticities that can be perceived by a standard observer; it can be seen as the result of performing this operation: slicing the XYZ volume with the plane $X+Y+Z=1$, then projecting the resulting plane onto the XY plane. See Figure 1.8.

It is worth remarking that the triplet of values formed by chromaticity (x, y) and luminance Y perfectly describes a color, and from (x, y, Y) we can obtain (X, Y, Z), and also (R, G, B).

The upper boundary of the chromaticity diagram is a horseshoe-shaped curve corresponding to monochromatic colors: this curve is called the spectrum locus [341]. The lower boundary is the purple line, and corresponds to mixtures of lights from the extrema of the spectrum. The relationship between the spectrum $\mathbf{E}$ of a light stimulus and its tristimulus values (X, Y, Z) is linear, given by the multiplication of the radiance by a matrix $\mathcal{S}_{\mathcal{XYZ}}$ whose rows are the color matching functions for the CIE XYZ color space:

$$\begin{bmatrix} X \\ Y \\ Z \end{bmatrix} = \mathcal{S}_{\mathcal{XYZ}}\mathbf{E}. \tag{1.18}$$

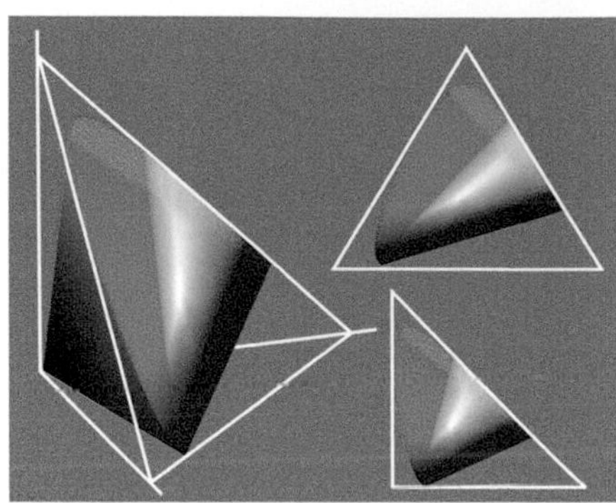

FIGURE 1.8: Left: XYZ volume. Top right: after slicing volume with plane $X + Y + Z = 1$. Bottom right: after projecting plane onto XY plane.

The linear relationship stated in Equation 1.18, combined with the definition of chromaticity coordinates in Equation 1.17, tell us the following: if monochromatic lights E_1 and E_2 have coordinates (x_1, y_1) and (x_2, y_2) (that will lie on the spectrum locus because the lights are monochromatic), the mixture $E_3 = E_1 + E_2$ will have coordinates (x_3, y_3) located in the segment joining (x_1, y_1) and (x_2, y_2). Therefore, the tongue-shaped region delimited by the spectrum locus and the purple line represents all the possible chromaticities that we can perceive, as mentioned above.

Mollon [276] explains how in 1852 Hermann Helmholtz provided a formal explanation for the differences between additive color mixing (of lights) and subtractive color mixing (of colored materials): for example, when we mix blue and yellow pigments we obtain green, but when we mix blue and yellow light we obtain white. Helmholtz suggested that pigments were composed of particles that absorbed light of some wavelengths and reflected some others, and the color of the pigment mixture will correspond to the wavelengths that are not absorbed by either of the constituents' pigments. In the aforementioned example, the yellow pigment reflects yellow, red and green but absorbs blue and violet, whereas the blue pigment reflects blue, violet and green and absorbs yellow and red; therefore, the light reflected from the mixture will be green. Helmholtz also showed, after a theoretical work by Hermann Grassmann (1853) that each monochromatic light had a complementary, i.e. the mix of both lights yields white. Monochromatic lights with wavelengths in the range between red and yellow-green have monochromatic complementaries with wavelengths in the range between blue-green and violets. The complementary of green is not a monochromatic light but purple, a mixture of blue and red light from the two ends of the visible spectrum.

Perfect white (i.e. light with a completely uniform power spectrum) has coordinates $x = y = \frac{1}{3}$, so as we mix a monochromatic light with white, its chromaticity coordinates move inwards and the saturation of the colors is reduced. A pure monochromatic light has 100% saturation while white has 0% saturation. But in practice, white lights never have a completely flat spectrum. The CIE has defined a set of standard illuminants: A for incandescent light, B for sunlight, C for average daylight, D for phases of daylight, E is the equal-energy illuminant, while illuminants F represent fluorescent lamps of various compositions [20]. The illuminants in the D series are defined simply by denoting the temperature in Kelvin degrees of the black-body radiator whose power spectrum is closer to that of the illuminant. A black-body radiator is an object that does not reflect light and emits radiation, and the power spectrum of this radiation is uniquely described by the temperature of the object. See Figure 1.9. For instance, CIE illuminant D65 corresponds to the phase of daylight with a power spectrum close to that of a black-body radiatior at $6500K$, and D50 to $5000K$. These two are the most common illuminants used in colorimetry [341]. The chromaticity coordinates of these illuminants form a curve called *Planckian locus*, shown in Figure 1.10. This is why in photography it is common to express the tonality of an illuminant by its *color temperature*: a bluish white will have a high color temperature, whereas a reddish-white will have a lower color temperature.

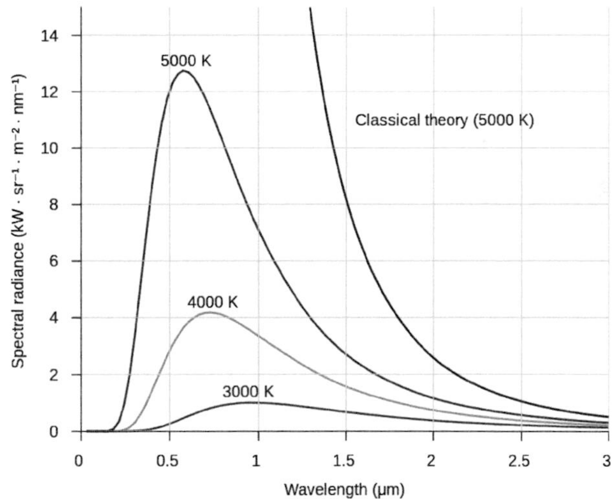

FIGURE 1.9: Spectrum for black-body radiators is a function of their temperature. Figure from [9].

Another very important consequence of the linearity property stated before is that any system that uses three primaries to represent colors will only be capable of representing the chromaticities lying inside the triangle determined by the chromaticities of the primaries. Furthermore, because of the convex shape of the chromaticity diagram, any such triangle will be fully contained in the diagram, leaving out chromaticity points. Therefore, for any trichro-

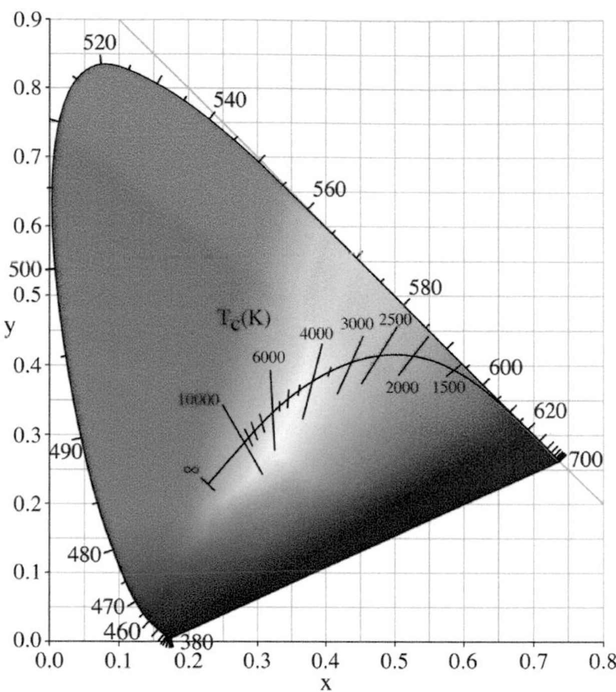

FIGURE 1.10: Chromaticity coordinates for black-body radiators as a function of their temperature. Figure from [9].

matic system there will always be colors that the system is not capable of representing. For instance, Figure 1.11 shows the chromaticity diagram for a cathode ray tube (CRT) television set, where the primaries are given by the spectral characteristics of the light emitted by the red, green and blue phosphors used in CRT's.

1.4 Perceptual color spaces

1.4.1 Color constancy and the von Kries coefficient law

Returning to Equation 1.1, if we light an illuminant of power spectrum $I(\lambda)$ on a surface of reflectance $R(\lambda)$, we receive from it a radiance $E(\lambda)$. Hence, a white sheet of paper illuminated by sunlight will produce radiance with a flat spectrum, which we should perceive as white, whereas the same

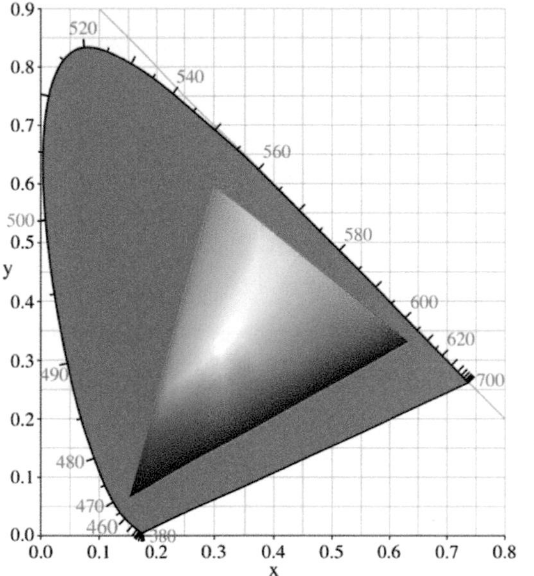

FIGURE 1.11: Chromaticities of a CRT television set. Image from [12].

paper under an orange light will produce radiance with more power in the longer wavelengths, which in theory we should perceive as orange. But as we know this is *not* what happens, we perceive the paper as being white in both cases, a manifestation of what is called *color constancy*, which is our ability to perceive objects as having a constant color despite changes in the color of the illuminant. In the example above, the *stimulus* is different (white light in the former case, orange light in the latter) but the *perception* is the same. Color constancy was already known in the late 17th century but it was first formally addressed by Gaspard Monge in 1789 [276]. Monge conjectured that our perception of color does not depend solely on the characteristics of the light reaching our eye; we also consider the color of the illuminant that lights the scene and we are able to "discount it." For instance, if an object under a red illuminant reflects white light to our eyes, we will perceive the object as being green, not white, because a white object would send red light. Or, in other words, if we receive white light and take away from it the red of the illuminant, we end up with green.

In 1905, Johannes von Kries formulated an explanation for color constancy that is known as von Kries' coefficient law and which is still used to this day in digital cameras to perform white balance. This law states that the neural response of each type of cone is attenuated by a gain factor that depends on the ambient light [344]. In practice, von Kries' law is applied by dividing each element of the tristimulus value by a constant depending on the scene conditions but not on the stimulus: typically, each element is divided by the corresponding element of the tristimulus value of the scene illuminant. Regardless of the original chromaticity of the illuminant, after applying the von Kries' rule the chromaticity coordinates of the illuminant become $x = y = z = \frac{1}{3}$, which correspond to achromatic, white light. In other words, von Kries' coefficient law is a very simple way to modify the chromaticity coordinates so that, in many situations, they correspond more closely to the *perception* of color.

1.4.2 Perceptually uniform color spaces

The CIE XYZ color space, and consequently the CIE xy chromaticity diagram, suffer from several limitations in terms of the perception of colors:

- the distance between two points in XYZ space or in the xy diagram is not proportional to the perceived difference between the colors corresponding to the points;
- a mixture of two lights in equal proportions will have chromaticity coordinates that do not lie exactly at the middle of the segment joining the chromaticities of the original two lights.

This can be observed in Figure 1.12, where equal-radius circles representing perceptual differences of the same magnitude are mapped into different-sized ellipses in the CIE xy diagram. Therefore we can say that, in terms of perception, the CIE XYZ space is not *uniform*.

But in color reproduction systems, perceptual uniformity is a very useful property because it allows us to define error tolerances, and therefore much work has been devoted to the developing of uniform color spaces. Research was carried out independently in two lines: finding a uniform lightness scale and devising a uniform chromaticity diagram for colors of constant lightness [341].

Lightness is the perceived level of light relative to light from a region that appears white, whereas brightness is the overall level of perceived light [344]. Experimentally it has been found that lightness is approximately proportional to the luminance raised to the power of $\frac{1}{3}$ [386].

In 1976 the CIE introduced two new color spaces: CIE 1976 $L^*u^*v^*$ (abbreviated CIELUV) and CIE 1976 $L^*a^*b^*$ (abbreviated CIELAB). They are both designed to be perceptually uniform, using the same $\frac{1}{3}$ power law for lightness L^* and different criteria for chromaticity. In CIELUV the chromaticity coordinates u^*, v^* are chosen so that just noticeably different colors are roughly equi-spaced. In CIELAB the chromaticity coordinates a^*, b^* are chosen so that the Euclidean distance between two points in CIELAB space is proportional to the perceptual difference between the colors corresponding to those points. Both CIELAB and CIELUV perform a normalization with respect to the tristimulus of a reference white, in what is a crude approximation to the color constancy property of the visual system; in CIELAB, it is directly based on the von Kries' coefficient law [341].

For CIELUV, the transformation from (X, Y, Z) to (L^*, u^*, v^*) coordinates is computed as

$$
\begin{aligned}
L^* &= \begin{cases} 116\left(\frac{Y}{Y_n}\right)^{\frac{1}{3}} - 16, & \text{if } \frac{Y}{Y_n} > \left(\frac{6}{29}\right)^3 \\ \left(\frac{29}{3}\right)^3 \frac{Y}{Y_n}, & \text{otherwise} \end{cases} \\
u^* &= 13L^*(u' - u'_n) \\
v^* &= 13L^*(v' - v'_n) \\
u' &= \frac{4X}{X + 15Y + 3Z} \\
v' &= \frac{9Y}{X + 15Y + 3Z} \\
u'_n &= \frac{4X_n}{X_n + 15Y_n + 3Z_n} \\
v'_n &= \frac{9Y_n}{X_n + 15Y_n + 3Z_n},
\end{aligned}
\tag{1.19}
$$

where (X_n, Y_n, Z_n) are the tristimulus values of a reference white, typically called the "white point" and which is usually taken as the brightest stimulus in the field of view: again, following von Kries' approach for color constancy; u'_n and v'_n are the (u', v') chromaticity coordinates of the white point. Figure 1.13 compares the (u', v') chromaticity coordinates with the xy chromaticity diagram.

In CIELAB, the lightness coordinate L^* is the same as in CIELUV and the chromaticities are computed as

$$
\begin{aligned}
a^* &= 500\left(f\left(\frac{X}{X_n}\right) - f\left(\frac{Y}{Y_n}\right)\right) \\
b^* &= 200\left(f\left(\frac{Y}{Y_n}\right) - f\left(\frac{Z}{Z_n}\right)\right),
\end{aligned}
\tag{1.20}
$$

where

$$
f(x) = \begin{cases} x^{\frac{1}{3}}, & \text{if } x > \left(\frac{6}{29}\right)^3 \\ \frac{1}{3}\left(\frac{29}{6}\right)^2 x + \frac{4}{29}, & \text{otherwise,} \end{cases}
\tag{1.21}
$$

which is the same power-law function used for the computation of the lightness L^*.

In CIELAB the chromaticity coordinates (a^*, b^*) can be positive or negative: $a^* > 0$ indicates redness, $a^* < 0$ greenness, $b^* > 0$ yellowness and $b^* < 0$ blueness. For this reason it is often more convenient to express CIELAB colors in cylindrical coordinates $L^*C^*h^*$, where

– $C^* = \sqrt{a^{*2} + b^{*2}}$, the radius from the origin, is the *chroma*, which can be defined as the degree of colorfulness with respect to a white color of the same brightness: decreasing C^* the colors become muted and approach gray;

- $h^* = arctan\frac{b^*}{a^*}$, the angle from the positive a^* axis, is the *hue*: a hue angle of $h* = 0°$ corresponds to red, $h* = 60°$ corresponds to yellow, $h* = 120°$ corresponds to green, etc.

Analogous correlates for chroma and hue exist for CIELUV in cylindrical coordinates: in that case, $C_{uv}{}^* = \sqrt{u^{*2} + v^{*2}}$ and $h_{uv}{}^* = arctan\frac{v^*}{u^*}$. Figure 1.14 shows the CIELAB color space in both Cartesian and cylindrical coordinates.

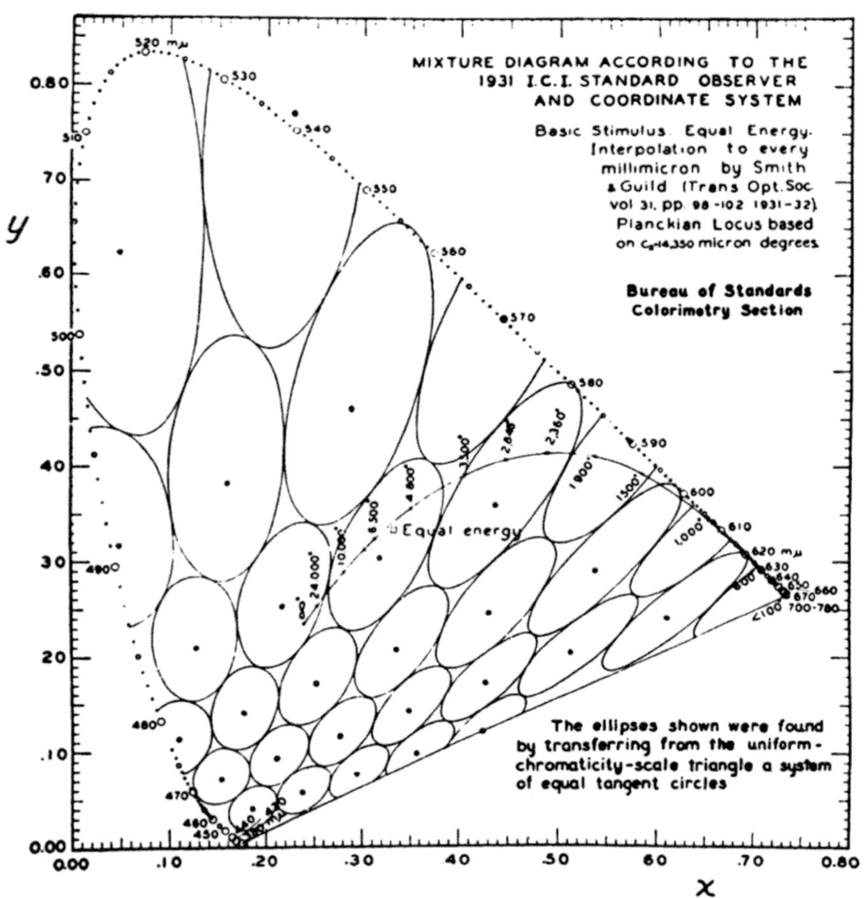

FIGURE 1.12: Ellipses representing the chromaticities of circles of equal size and constant perceptual distances from their center points. Figure from [215].

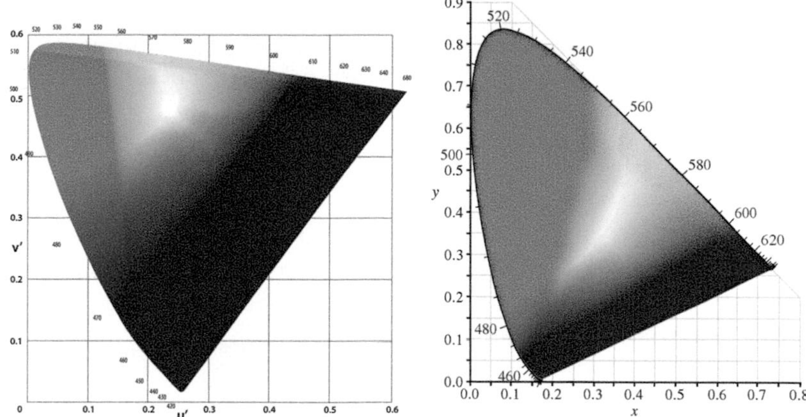

FIGURE 1.13: Left: (u', v') chromaticity coordinates, from [11]. Right: (x, y) chromaticity coordinates from the CIE 1931 xy chromaticity diagram. Figure from [13].

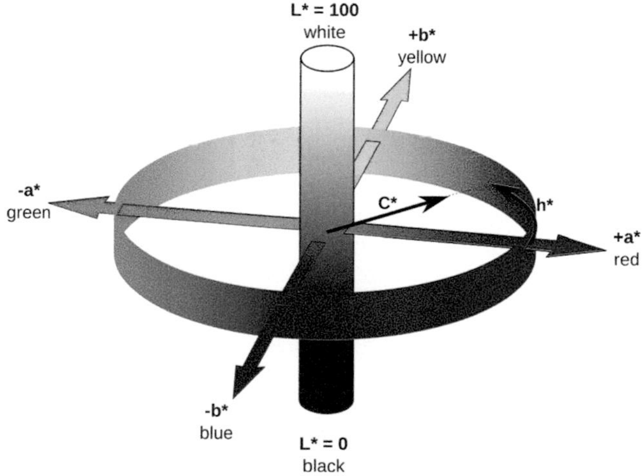

FIGURE 1.14: CIELAB color space in both Cartesian and cylindrical coordinates.

1.4.3 Limitations of CIELUV and CIELAB

We mentioned that in the CIE xy diagram, Euclidean distances between chromaticities do not correspond to perceptual differences between colors, e.g. the mixture of two colors in the same proportion should be located at the midpoint of the segment determined by these colors in the diagram, but generally this is not the case. And this motivated the research on uniform color spaces CIELAB and CIELUV, where if the Euclidean distance between a pair of colors is the same as the distance between another pair, then the corresponding differences in perception are also (roughly) equal. But this is just an approximation, and CIELUV and CIELAB are both only partially uniform.

In some parts of the color space (mainly around blue) CIELAB suffers from *cross-contamination* [282]: changing only one attribute (such as hue) produces changes in the perception of another attribute (such as saturation). This is a consequence of the deficiencies of the system with respect to the correlates for hue [341]: the correlate for hue is the angle $arctan\frac{b^*}{a^*}$, and therefore constant hue should correspond to planes passing through the L^* axis, but what is observed experimentally are curved surfaces instead of planes. These surfaces depart more from the intended planes near the negative b^* axis, hence the problems around blue.

CIELUV also suffers from approximate perceptual uniformity and poor hue correlation in some regions; furthermore, the translational way of imposing white point normalization (i.e. computing differences $u' - u'_n$ and $v' - v'_n$) may create imaginary colors (falling outside the spectral locus) in some contexts. Shevell [344] explains that when the CIE introduced CIELUV and CIELAB in 1976 the organization recognized that it was difficult to choose one color space over the other because each worked better on different data sets, but more recently the general opinion seems to favor CIELAB, and CIELUV is no longer widely recommended [156].

Another, very important, limitation of both color spaces is that they are only useful for comparing stimuli under similar conditions of adaptation [341]: they don't consider any of the factors affecting color appearance in real-life conditions.

1.5 Color appearance

In laboratory conditions, the three perceptual dimensions of hue, saturation and brightness characterize how we perceive the colors of single, isolated lights. But in real-life situations, lights are seldom isolated and our perception of the color of an object is not completely determined by the light coming from it; it is influenced by factors such as the ambient illumination, the light coming from other objects, or the current state of the neural pathways in eye and brain [344].

We have insisted upon the fact that color appearance is a perceptual, not physical, phenomenon, and have seen an example in the color constancy property of the human visual system: the color appearance of objects is in good correspondence with their reflectance, despite changes in the illuminant, i.e. the light reaching our eyes changes but our perception remains the same. Another example is that of color induction: objects that send the same light to our eyes are perceived as having different colors because of the influence of their surroundings. Figure 1.15 shows an example: the inner rings in both sets of concentric rings are identical, as the square in the middle shows, yet they appear to have very different hues.

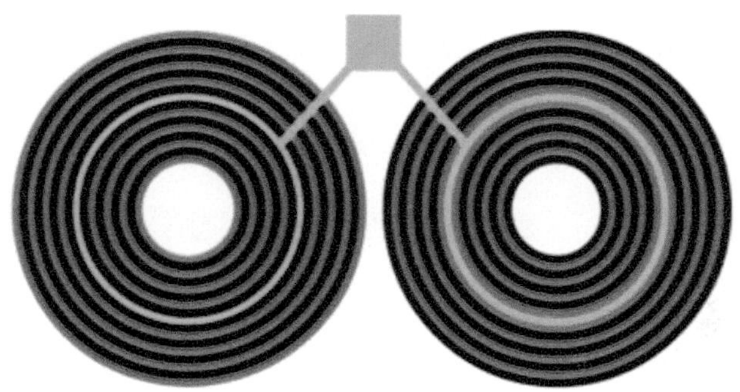

FIGURE 1.15: Chromatic induction: color appearance depends on surroundings. The inner rings are identical, yet they appear to us as having different colors. Figure by P. Monnier [277].

Figure 1.16 shows the famous checker board illusion by Edward Adelson. Squares A and B are identical (as the image at the right shows) but in the left image they appear to have very different lightness.

FIGURE 1.16: The "checker shadow illusion" shows that lightness depends on context. Figure by Edward H. Adelson (1995), from [10].

Likewise, our perception of saturation and contrast is also dependent on ambient conditions: saturation decreases when the illumination decreases (the Hunt effect), contrast also decreases with diminishing illumination (the Stevens effect). See Figure 1.17.

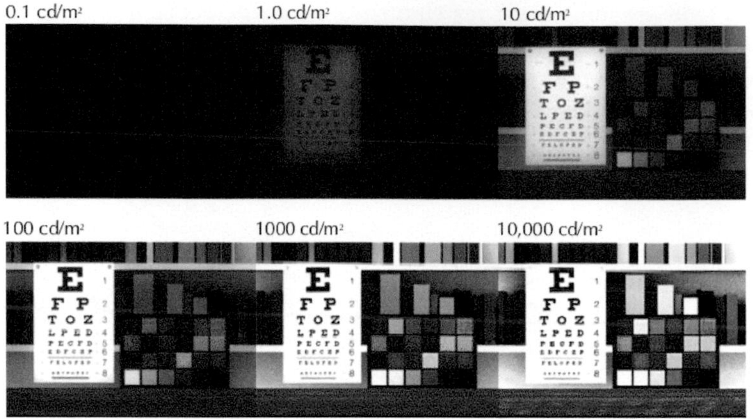

FIGURE 1.17: Hunt and Stevens effects. Image from [155].

Neither the CIEXYZ color space nor CIELUV or CIELAB are useful to compare stimuli under different adaptation conditions, for this we don't need a color space but a *color appearance model.* For given colorimetry under specified reference viewing conditions, a color appearance model predicts the colorimetry required under the specified test viewing conditions for producing the same color appearance [341].

The CIE has proposed several color appearance models, the most recent of which is CIECAM02, published in 2002. The two major pieces of CIECAM02 are a chromatic adaptation transform and equations for computing correlates of perceptual attributes, such as brightness, lightness, chroma, saturation, colorfulness and hue [281]. The chromatic adaptation transform considers chromaticity variations in the adopted white point and was derived based on experimental data from corresponding colors data sets. It is followed by a non-linear compression stage, based on physiological data, before computing perceptual attributes correlates. The perceptual attribute correlates were derived considering data sets such as the Munsell Book of Color. CIECAM02 is constrained to be invertible in closed form and to take into account a sub-set of color appearance phenomena.

Chapter 2

Optics

2.1 Introduction

A basic understanding of the physical properties of light and its interaction with the optical elements present in cameras is essential in order to identify the many kinds of visual artifacts that may appear in recorded images, as well as to compensate them or reduce them as much as possible with image processing techniques.

In this chapter we will begin by discussing visible light as a form of radiation, and introduce ray diagrams as a very convenient form of representation of light trajectories. Next we will discuss specular and diffuse reflection and refraction, which will allow us to explain how lenses work and which image problems they may cause, which are called optical aberrations. Then we will introduce several important terms in photography, such as angle of view, depth of field and f-number, and relate them to different lens categories. Finally we will discuss the Modulation Transfer Function (MTF) and its relationship with image resolution and the diffraction of light.

2.2 Ray diagrams

Instead of modeling light as waves (which would be useful to describe its color) or particles (something we would need to do if, for instance, we were to deal with interactions at the atomic level), a much more simple but very powerful model consists of using ray diagrams to represent how light travels through space [134]. In a ray diagram, the trajectory of a ray of light is represented with a straight line ending in an arrow-head: this is the direction of propagation of the lightwave, and hence it's perpendicular to the wavefront. Ray diagrams allow us to grasp visually the behavior of light when it interacts with materials, without making computations. Here follow some guidelines, taken from [134], for a proper use of ray diagrams.

In Figure 2.1 the light rays change direction when they go from the water

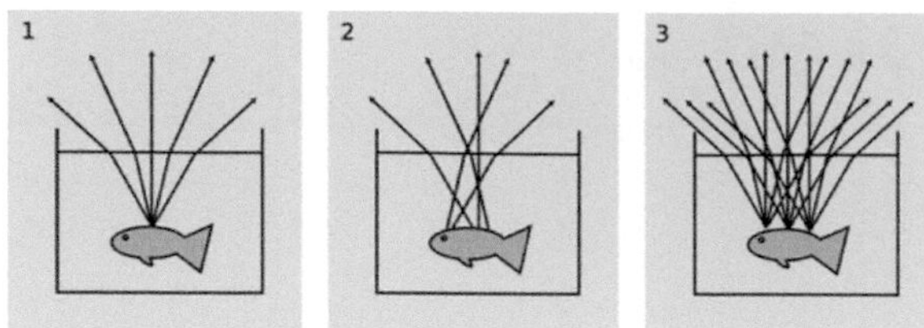

FIGURE 2.1: The use of ray diagrams, figure taken from [134]. 1. Correct. 2. Incorrect: implies that diffuse reflection only gives one ray from each reflecting point. 3. Correct, but unnecessarily complicated.

into the air. To an observer in the air looking down onto the water, these rays seem to come from a point that is not on the fish but above it.

In Figure 2.1.1 there are several rays coming from the same point. This is what happens with *diffuse* reflection, where the light reaching a diffuse surface such as the fish's is reflected in infinite directions (of which the diagram only represents five, for clarity). While light rays may be of infinite length, we have to decide when to start and when to end the lines in a ray diagram. If we want the ray diagram for this example to tell us where the fish will be seen from above the water, then it makes sense to start the ray lines at a point on the surface of the fish, and *not* to draw how light got to that point before being reflected (and since the reflection is diffuse, the direction of light incident on the fish would tell us nothing.) It also makes sense to end the lines once they are out of the water, since we would gain nothing by making them longer than what they are now.

Figure 2.1.2 is incorrect in the sense that it is a confusing ray diagram, which implies that each point on the fish's surface only reflects light in one direction. We would not be able to predict, using this ray diagram, where the fish is seen from above water.

Figure 2.1.3 is incorrect in the sense that it detracts from the clarity of a ray diagram like the one in Figure 2.1.1. In this example, all points on the surface of the fish behave in the same way so we only need to show what happens in one; adding more points to the diagram only confuses the picture.

2.3 Reflection and refraction

When light reaches an object, one or more of these phenomena will happen:

- it will be reflected, in one direction (specular reflection) or many directions (diffuse reflection);

- it will be transmitted through the object, changing the direction of its path, in a phenomenon called refraction;
- it will be absorbed, heating up the object.

In the case of specular reflection, the angle (with respect to the surface normal) of the reflected ray will be the same as that of the incident ray; furthermore, it will be in the same plane as the one formed by the surface normal and the incident ray. Diffuse reflection, as we mentioned, happens in infinite directions, although depending on the material properties it may be more significant in some directions than in others. See Figure 2.2 for a ray diagram.

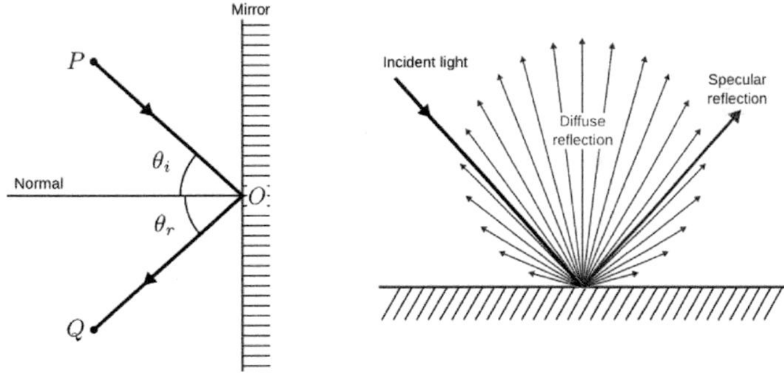

FIGURE 2.2: Types of reflection: specular (left), diffuse (right).

In the case of refraction, if light travels from medium 1 with refraction index n_1 to medium 2 with refraction index n_2, the behavior of light rays at the interface of both mediums is modeled by Snell's law which can be stated as:

$$n_1 sin\theta_1 = n_2 sin\theta_2, \tag{2.1}$$

where $\theta_i, i = 1, 2$, represent angles with respect to the interface normal. See Figure 2.3.

The index of refraction of a material is a unit-less value related to the density of the material (as density increases, so does the refraction index) and the velocity of light in said material. Its minimum possible value is 1, the value vacuum has. Air has an index of approximately 1.0003, while water's index is around 1.33.

While the index of refraction is constant for any given substance, it depends on the wavelength of the light, e.g. glass has different refraction indexes for lights of different colors. This, in conjuntion with Snell's law, explains why a glass prism is able to separate a ray of white light (which is composed of light of different wavelengths, covering the whole visible spectrum) into different rays for different colors, as in the famous experiment performed by Newton; see Figure 2.4.

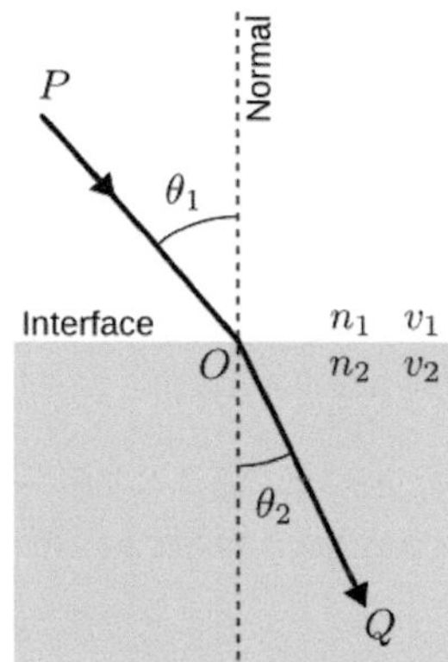

FIGURE 2.3: Refraction and Snell's law.

Snell's law as written in Equation 2.1 also tells us that for some values of n_1, n_2, θ_1 there isn't any possible value of θ_2 that satisfies the equality. Specifically, when $n_1 > n_2$ there always exists a critical angle θ_c such that when $\theta_1 > \theta_c$ there isn't any possible value for θ_2, because we would need an angle θ_2 with a *negative* sine in order to satisfy Snell's law. In these cases, there is *no* refraction and the light does not go into medium 2, it is reflected back into medium 1 in a phenomenon called *total internal reflection*. This is how fiber optics work, how endoscopies work: light rays travel inside the fiber, they reach its surface but never cross it (hence there's no refraction), they keep being reflected back into the fiber.

2.4 Lenses

Lenses are specially shaped pieces of glass or other material with refraction index substantially higher than that of air, with two surfaces that can be curved (of either concavity) or planar. Therefore, light rays that go through a lens are refracted twice, once on entering and again on leaving the lens, and the behavior of any lens can be described just by using Snell's law. The mathematical derivations in this section are adapted from the course notes of [61].

2.4.1 Refraction at a spherical surface

We start by considering the first refraction stage, when light enters the lens. Since the refraction index of the lens material is larger than that of air,

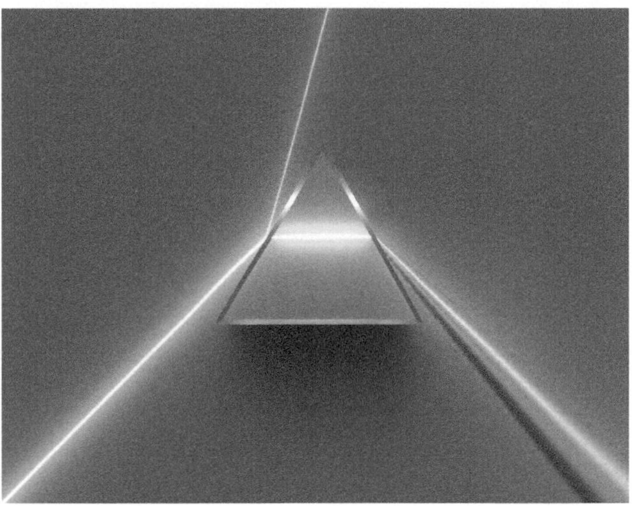

FIGURE 2.4: A prism separating white light into different colors, as observed by Isaac Newton. Image from [17].

there is no total internal reflection. Light rays that are parallel to the optical axis are focused by the spherical surface: we will prove that they all cross at a point, the focus, that lies on the optical axis at a distance f from the surface.

Figure 2.5 shows a convex spherical surface of radius R and center of curvature O, which is the interface between air (index of refraction = 1) and glass (index of refraction n). Rays parallel to the optical axis imply that all angles considered in the figure are small, therefore we can approximate them by their sine.

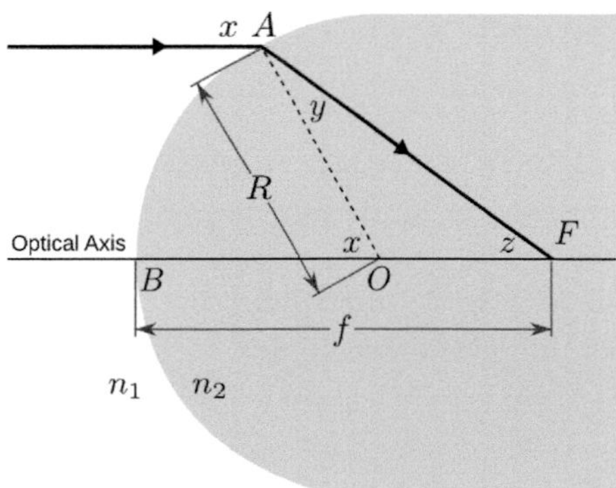

FIGURE 2.5: Focus for a convex boundary. Figure adapted from [61].

At point A, Snell's law can be written as $x = ny$, which implies $\frac{x}{y} = 1$. The triangle angle sum theorem for triangles ABO and AOF gives us the angle relationship $z = x - y$. The length of the arc AB is Rx, and since z is small we also have $AB \simeq fz = f(x - y)$.

From all the above:
$f \simeq \frac{AB}{x-y} = \frac{Rx}{x-y} = \frac{R\frac{x}{y}}{\frac{x}{y}-1} = \frac{Rn}{n-1}$,
that is, the focal distance f is a function of the curvature of the surface and of the refraction index of the lens, but is independent of the location of the point A that we used in the derivation. Therefore, all rays parallel to the optical axis are focused at distance f, as we wanted to prove. The same result can be shown for a concave boundary, only in that case the refracted rays don't cross at the focus, they seem to diverge from it (i.e. the prolongations of the refracted rays do cross at the focus).

Next, we can compute the precise location i of the image I of a source S placed at s; see Figure 2.6.

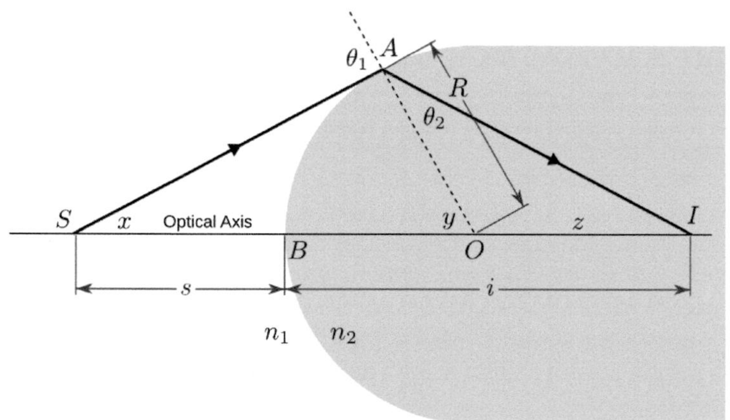

FIGURE 2.6: Location of an image. Figure adapted from [61].

From triangle ASO: $\theta_1 = x + y$, and from triangle AOI: $\theta_2 = y - z$. Therefore, with a small angle approximation, Snell's Law gives: $x + y = n(y - z)$.

The arc AB has length Ry in radians, but for small angles the following equations also hold (approximately): $AB = sx = iz$. Combining these with the above, we get:

$$\frac{1}{s} + \frac{n}{i} - \frac{n-1}{R}. \tag{2.2}$$

It turns out that this Equation also holds for concave boundaries, if we use the following sign conventions, as stated in [61]:

Let Side A of an optical component be the side from which light starts, and let Side B be the side to which light travels.

The sign of s is determined by Side A. If the source is on Side A, s is positive; if it is on the side opposite to Side A, s is negative.

The signs of i, f, R are determined by Side B. For the image and the focal point, their measurements are positive if they are on Side B, and negative if they are on the side opposite to Side B. R is positive provided the centre of curvature is on Side B; it will be negative if it is on the side opposite Side B.

2.4.2 The lens-maker's equation

As it is usually done, we will consider the *thin lens* approximation: we will assume that the lens has negligible thickness. This simplification is quite useful since it allows us to disregard how light travels inside the lens and only consider its refraction at the surfaces (also, it makes the lens symmetrical, i.e. the location of source, image and focus will not change if we flip the lens exchanging its sides).

Therefore, to compute the location i of an image I produced by a lens from an object S placed at s, we must apply Equation 2.2 twice, once for refraction at each surface of the lens, being careful with the sign conventions and exchanging the refraction indexes:

- The first surface, of radius R_1, forms image I_1 (located at i_1) from source S: $\frac{1}{s} + \frac{n}{i_1} = \frac{n-1}{R_1}$.
- The second surface, of radius R_2, forms image I from source I_1: $\frac{n}{-i_1} + \frac{1}{i} = \frac{1-n}{R_2}$. Notice how we have changed the sign of the source and exchanged indexes 1 and n since now light is travelling from the lens to the air.

Adding these two equalities we get the lens-maker's equation:

$$\frac{1}{s} + \frac{1}{i} = (n-1)\left(\frac{1}{R_1} - \frac{1}{R_2}\right). \tag{2.3}$$

The term on the right-hand side is a constant for the lens, depending only on its shape and refraction index. In fact, it's the inverse of the focal distance: when the source S is placed at infinity $\frac{1}{s}$ is zero, and since the rays are parallel to the optical axis the image is formed at the focus, located at f, so $i = f$ and therefore $\frac{1}{f} = (n-1)\left(\frac{1}{R_1} - \frac{1}{R_2}\right)$:

$$\frac{1}{s} + \frac{1}{i} = \frac{1}{f} = (n-1)\left(\frac{1}{R_1} - \frac{1}{R_2}\right). \tag{2.4}$$

When $f > 0$, as with a bi-convex lens, the lens is called convergent because it makes a parallel beam converge at the focus, whereas if $f < 0$ the lens is called divergent because it makes a parallel beam of rays diverge from the focus; see Figure 2.7.

Finally, notice how the lens-maker's Equation shows that the lens is symmetric: if we move the source to the other side of the lens and shine light from

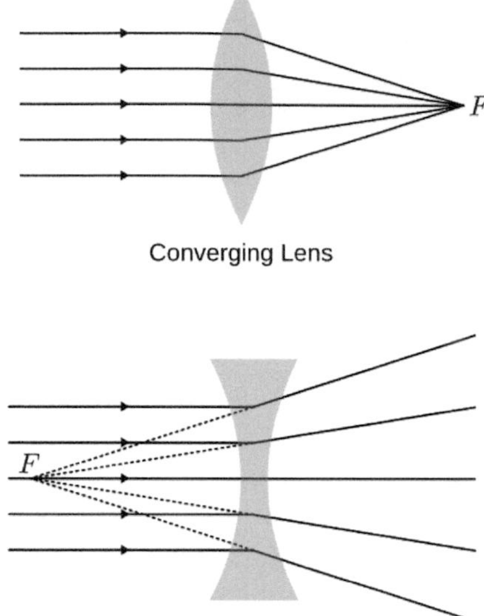

FIGURE 2.7: Converging and diverging lenses. Figure adapted from [61].

the source's side, the focus is exactly the same. (This shows that a lens has two foci, one on each side of the lens, with equal distance to the lens) [61].

2.4.3 Magnification

To compute the magnification of a lens we make use once more of the *thin lens* assumption, which allows us in this case to say that a light ray going through the center of the lens doesn't change its direction. The reason why is that with a neglible width, a lens is basically flat close to its center, and a flat piece of glass does not change the direction of the light that goes through it, according to Snell's law. Therefore, we have the situation depicted in Figure 2.8. We trace two rays from the source, one parallel to the axis that will be refracted and pass through the focus, and one through the lens' center, which will not change its course. Both rays cross at the image, and by similar triangles we get the magnification formula: $M = -\frac{i}{s}$, which, combined with the lens-maker's equation, gives:

$$M = \frac{f}{f - s}. \tag{2.5}$$

Notice that the convention is that inverted images have negative magnification, although the size of the image may be larger than that of the source (in which case the absolute value of the magnification is larger than one).

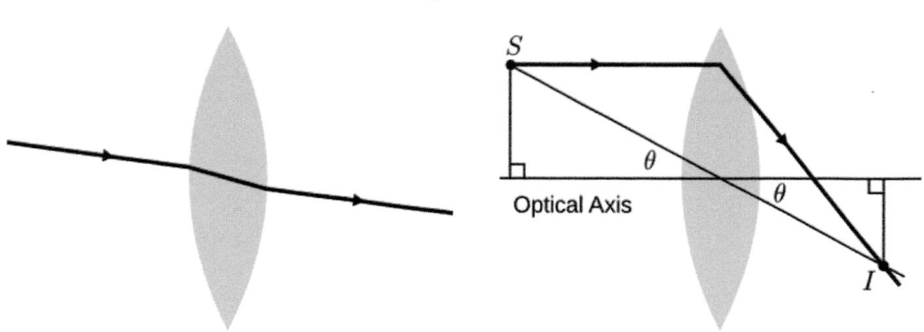

FIGURE 2.8: Magnification of a lens. Figure adapted from [61].

2.4.4 Lens behavior

For a diverging lens $f < 0$ and from Equations 2.4 and 2.5 we get that the images are always virtual ($i < 0$ so refracted rays do not actually cross), diminished (the absolute value of the magnification is less than one) and upright (the magnification has a positive sign, so the image is not inverted). See Figure 2.9.

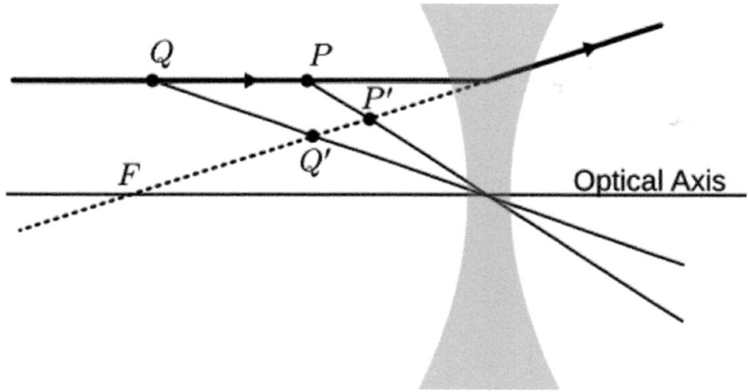

FIGURE 2.9: Image for diverging lens. Figure adapted from [61].

For a converging lens $f > 0$ and we have three cases:

- If $s > 2f$ then $i > 0, M < 0$ and $|M| < 1$ therefore the image is real, inverted and diminished; see Figure 2.10.
- If $2f > s > f$ then $i > 0, M < 0$ and $|M| > 1$ therefore the image is real, inverted and magnified; see Figure 2.11.
- If $f > s > 0$ then $i < 0, M > 0$ and $|M| > 1$ therefore the image is virtual, upright and magnified; see Figure 2.12.

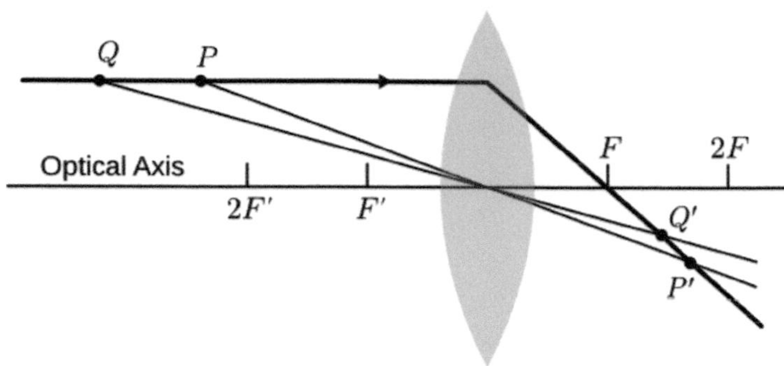

FIGURE 2.10: Converging lens, far away source. Figure adapted from [61].

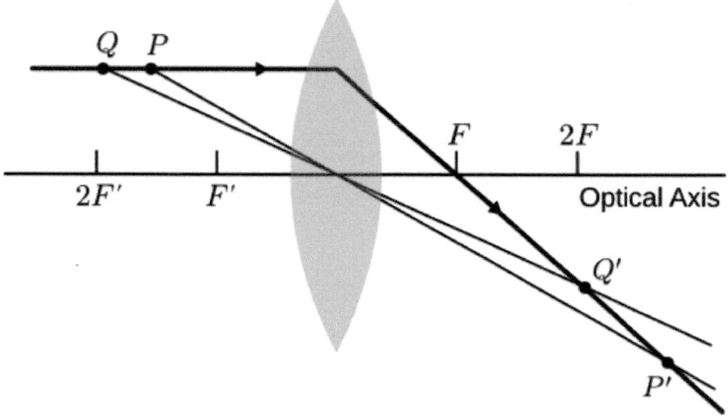

FIGURE 2.11: Converging lens, source between one and two focal lengths. Figure adapted from [61].

2.5 Optical aberrations

Because of the different approximations used in our modeling of a lens we obtain a very neat result: different rays from the same source point always get focused onto the same image point. In reality, this does not always happen and we get image problems called optical aberrations, where points don't appear as points, planes don't appear as planes, or subject and object have different shape (we will see some examples in Figure 2.19). Using polynomials to approximate lens surfaces, third-order errors cause aberrations that were classified by Seidel in 1856 and hence bear the name of Seidel aberrations [235]:

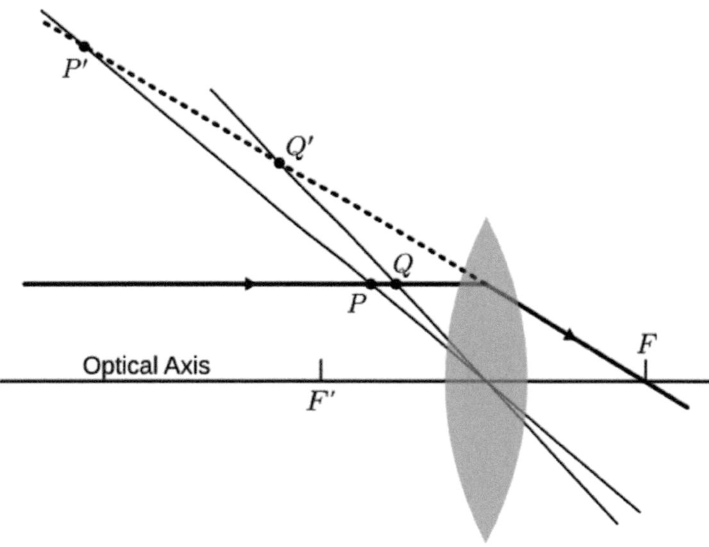

FIGURE 2.12: Converging lens, close source. Figure adapted from [61].

– *Spherical aberration.* The spherical shape of the lens causes rays that are far from the optical axis to be focused much closer to the lens than rays which come near the optical axis; see Figure 2.13.

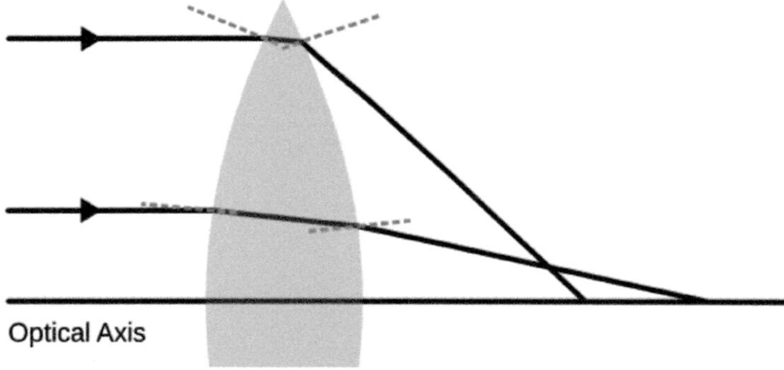

FIGURE 2.13: Spherical aberration. Figure adapted from [61].

As a result, point light sources near the axis have blurred images. Building aspherical lenses, which don't have this problem, is technically challenging. A common correction is to combine the spherical lens with another, with a certain refraction index so that both lenses in conjunction behave like an aspherical lens; see Figure 2.14.

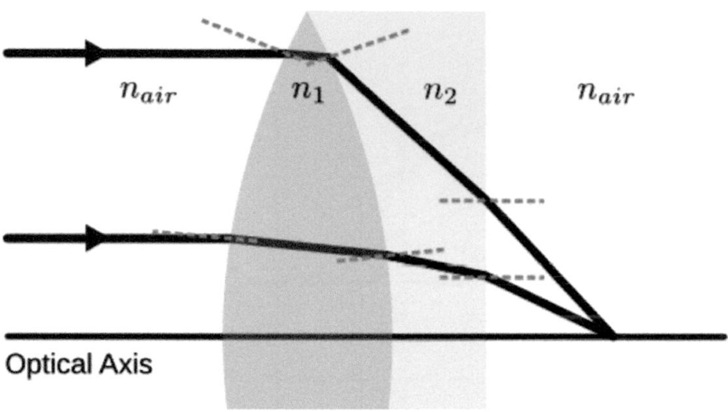

FIGURE 2.14: Fixing spherical aberration. Figure adapted from [61].

– *Comatic aberration.* Rays from a point source off the axis cross at several image locations, producing a comet-like flare, hence the name. This aberration is not present at the image center (unlike spherical aberration), but gets progressively worse as the source point gets further away from the axis; see Figure 2.15. Coma can be corrected by reducing aperture size, and also by adjoining a second lens.

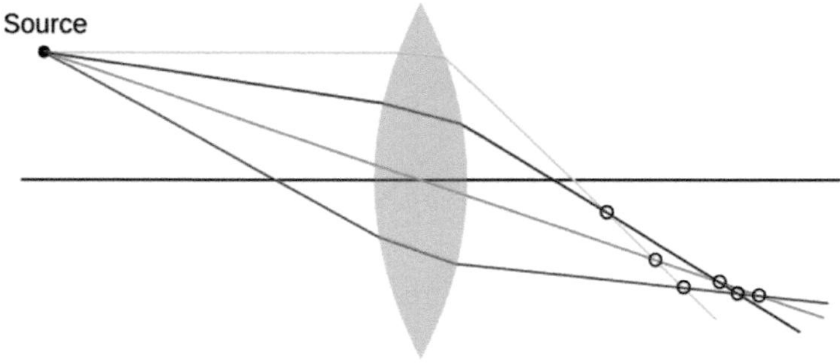

FIGURE 2.15: How coma forms. Figure adapted from [61].

– *Astigmatism* is a form of aberration that happens when the lens' surface has different curvature in different directions. From the lens-maker's Equation we can see that this implies different focal lengths in different directions, and point sources get projected as ellipses. A possible correction is to reduce the aperture.

– *Field Curvature.* With this type of aberration, a planar object looks curved and the periphery of the image is out of focus. It is related to spherical aberration, because it is also due to the fact that spherical lenses focus rays on different points depending on their distance from the optical axis; see Figure 2.16. Again, a possible correction is to reduce the aperture.

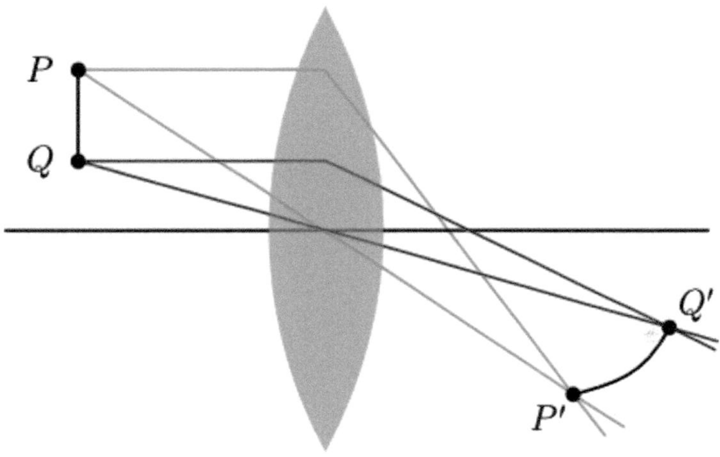

FIGURE 2.16: Field curvature. Figure adapted from [61].

- *Distortion.* In this form of aberration, images are distorted because the magnification is not constant over the image area. If magnification is larger at the image edges we get the *pincushion distortion* shown on the left of Figure 2.17, whereas if magnification decreases from the image center we get the *barrel distortion* shown on the right of Figure 2.17: in both cases, the source was a square but the image is curved. The reason for this type of aberration comes from Equation 2.5, which showed that the magnification depends on the focal distance f: since for spherical lenses f varies with the distance to the image center, so does the magnification.

The above listed five types of aberrations depend only on the geometry of the lens, to which we can add the problem of *vignetting*, in which the brightness of the image decays at the boundaries of the frame. In the case of optical vignetting the problem is caused by the physical length of the lens system, which prevents light off the optical axis from reaching the final lens element and/or the camera sensor. There are other kinds of vignetting as well, and digital cameras suffer from pixel vignetting as we'll see in Chapter 3.

There is also another, very important kind of aberration that is due to the chromaticity of the light:

- *Chromatic aberration.* It can be axial or lateral. Axial or longitudinal chromatic aberration is due to the fact that the index of refraction varies

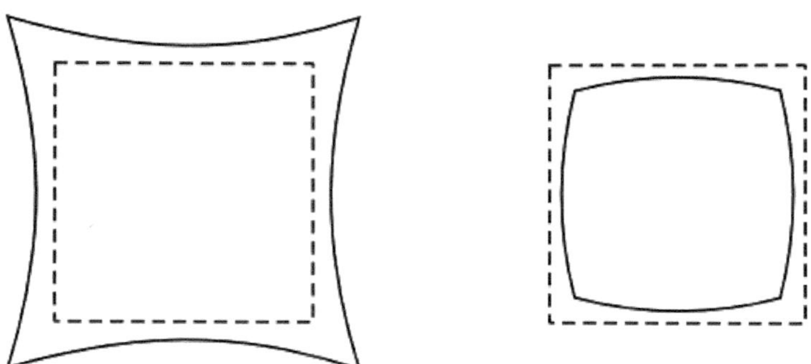

FIGURE 2.17: Distortion. Figure adapted from [61].

with wavelength, with the net effect that blue light gets focused closer to the lens than red light; see Figure 2.18.

Axial chromatic aberration therefore can be reduced by diminishing the aperture size. On the other hand, lateral chromatic aberration is due to different colors having different rates of magnification, causing point-symmetrical color-bleeding from the center towards the margins of the image [235], and aperture reduction cannot solve this problem.

As explained in [61], chromatic aberration is usually tackled by joining lenses together so that their aberrations cancel each other, but unfortunately it is very difficult to remove because the configuration of lenses that makes light of two colors to be focused on the same point does not usually work with some other two different colors.

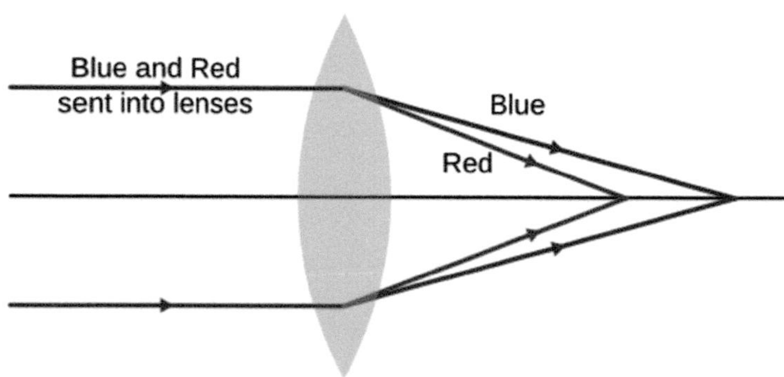

FIGURE 2.18: Chromatic aberration. Figure adapted from [61].

FIGURE 2.19: Optical aberrations. From left to right and top to bottom: spherical aberration, field curvature, coma (notice the star appears as a comet), astigmatism, pincushion distortion, barrel distortion, chromatic aberration (notice edge of house on the right), vignetting. Images from [41, 39, 45, 47, 53, 44, 48, 54].

2.6 Basic terms in photography

In order to present some basic definitions of optical properties of cameras and lenses, we first need to introduce a very simple camera model. More details on cameras will be given in the next chapter.

Figure 2.20 shows the simplest possible schematic of a camera. When the shutter is open, light passes through the lens and reaches the sensor, where it is recorded; images are inverted. Camera lenses are actually lens *systems*, that we treat and think of as a single lens with a given focal distance, but which in reality are a combination of lens elements: with several lens elements we can correct aberrations, and allow for varying focal distance.

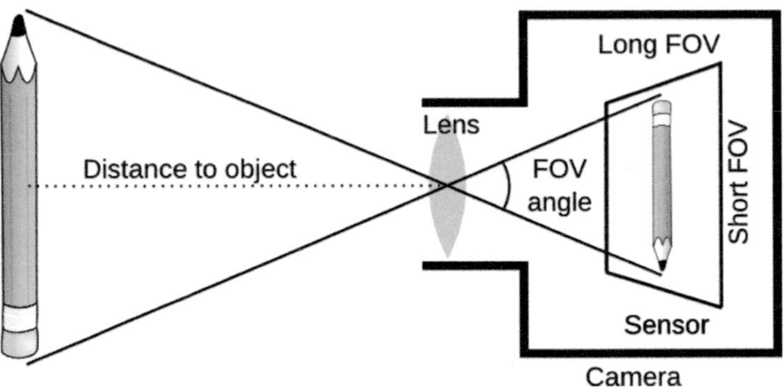

FIGURE 2.20: Basic camera diagram.

The shutter speed or exposure time is the amount of time the shutter is open, usually measured in fractions of a second, e.g. 1/200.

The aperture is the area of the lens that lets the light go through. Its size is regulated by the diaphragm, a mechanism where moving plates change the lens opening, as seen in Figure 2.21. Photographers do not operate directly on the aperture, which could be measured in terms of area or diameter, but prefer instead to use the *relative aperture* or f-number, a dimensionless unit which we will discuss shortly.

Given the focal distance f of the lens and the position s of an object S, the lens-maker's Equation tells us where the image will be formed, at location $i = \frac{fs}{s-f}$. The image plane is the plane where the sensor is located: if we put the image plane at a distance i from the lens, then the object S will appear sharp in the image. If the image plane is at any other distance, then S will appear blurred; see Figure 2.22. Therefore, to focus on any given object the distance between the lens and the sensor must be adjusted.

The angle of view is the angle enclosing the largest object whose image

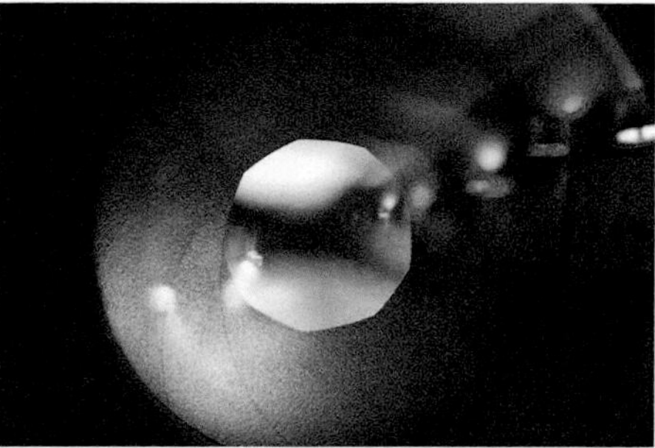

FIGURE 2.21: Iris diaphragm with nine blades. When the blades move, the aperture changes. Image from [56].

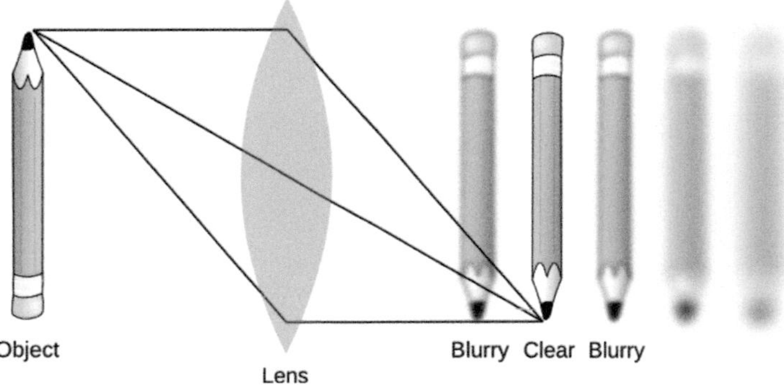

FIGURE 2.22: Focus is regulated varying the lens distance to the image plane. Image adapted from [6].

can fit on the sensor. In Figure 2.23 the sensor has size d and the angle of view is α.

The angle of view is determined by the focal length of the lens. Lenses with a focal length producing a field of view similar to the one we perceive with our naked eye are called normal lenses. If the focal length is short the angle of view will be wide and the lens is hence called a wide-angle lens. If the focal length is long the angle of view is small and the lens is called long lens, telephoto lens or simply tele lens. The angle of view also has a direct effect on perspective: a wide-angle lens makes the background appear much distant and smaller, whereas a long lens brings the subject and the background closer. See Figure 2.24.

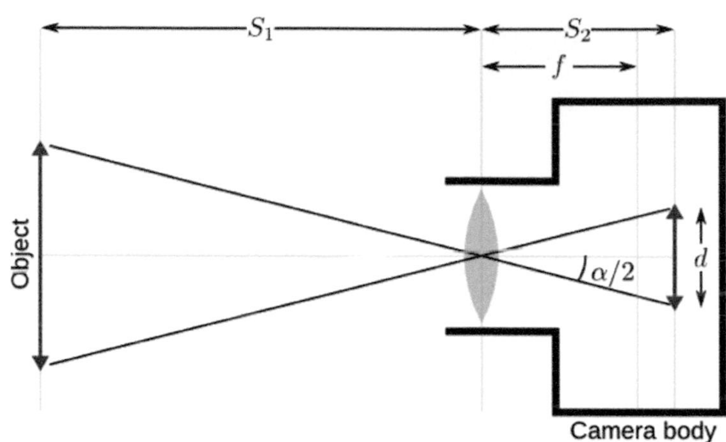

FIGURE 2.23: Angle of view.

FIGURE 2.24: Increasing the focal length, the angle of view decreases.

2.6.1 f-number

The f-number of a lens is a dimensionless quantity that expresses the lens *speed*, which is a measure of the amount of light a lens needs in order to produce a bright-enough image (a fast lens needs less light than a slow one).

Before we can define it, we will introduce some other concepts and make some computations following [27]. We have a small object source of brightness B and area A_s at distance s from the lens, which produces an image of area A_i located at i; the lens has area A and diameter D. We can compute the power of the light from the source that passes through the lens: it is the product of the brightness B times the object area A_s times the solid angle subtended by the aperture: $B \cdot A_s \cdot \frac{A}{s^2}$. Then, the illuminance E or power arriving per unit area of the image is $E = B \cdot A_s \cdot \frac{A}{s^2} \cdot \frac{1}{A_i}$.

The magnification M is $M = \frac{i}{s}$ and therefore $M^2 = \frac{A_i}{A_s}$. With the lens-maker's Equation $\frac{1}{s} + \frac{1}{i} = \frac{1}{f}$ and substituting $A = \frac{\pi D^2}{4}$, we get that the illuminance is

$$E = \frac{\pi B}{4N^2}, \tag{2.6}$$

where

$$N = (M+1)\frac{f}{D} \tag{2.7}$$

is the f-number of the lens. Typically, for objects far away from the lens, $M \simeq 0$ and the f-number becomes

$$N = \frac{f}{D}, \tag{2.8}$$

which is the usual way of expressing it (for macro photography M is not negligible and Equation 2.8 must be adapted). Given that D is the aperture diameter, the f-number is often called *relative aperture*. The notation is *f/N*, e.g. a lens with aperture diameter $D = 25mm$ and focal length $f = 50mm$ will have an f-number of $f/2.0$. Equation 2.8 gives the incorrect impression that by increasing the size of the lens we may get an arbitrary small value for N, but the theoretical minimum is actually 0.5: Figure 2.25 shows that $f \cdot tan(\theta) = D/2$, for small angles we have $tan(\theta) \simeq sin(\theta)$, therefore $\frac{f}{D} = \frac{1}{2sin(\theta)}$ and the f-number is always larger than $\frac{1}{2}$. In practice the fastest lens on the market has $N \simeq 1$ while many lenses have $N \simeq 2$ [235].

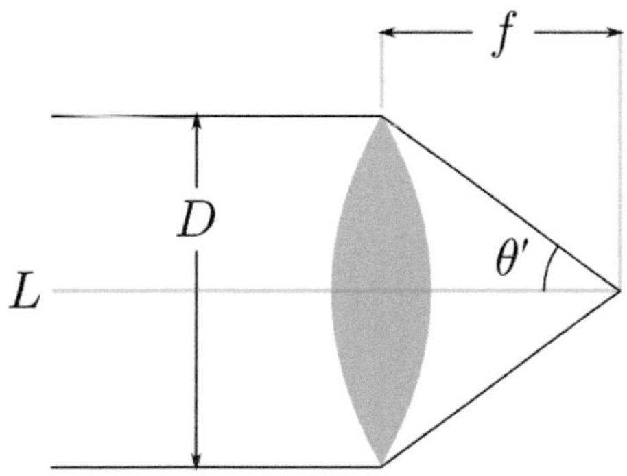

FIGURE 2.25: Schematic of single lens, adapted from [235].

At a fixed focal length f, decreasing the aperture size D increases the f-number N. By halving the aperture diameter we are doubling the f-number, and hence the illuminance E is reduced by four, as Equation 2.6 shows. In practice, photographers find it useful to work with ratios of illuminance amounting to changes by a factor of two: given a lens configuration, they want to know how to modify it so that the amount of light reaching the image is doubled, or halved. Therefore, f-numbers are not allowed to vary continuously, but rather they come in a sequence where from one value to the next the illuminance is halved. Because of the quadratic term in the f-number in Equation 2.6, a sequence where E is halved at each step implies that N is multiplied by $\sqrt{2}$ at each step and this is why the standard scale for f-numbers is approximately the sequence of powers of the square root of 2: $f/1, f/1.4, f/2, f/2.8, f/4, f/5.6$, etc. Each increment in the f-number sequence is called *f-stop* or simply *stop*. For instance, going up two stops implies decreasing illuminance by four.

If two lenses with different focal lengths are used with the same f-number, then they are both letting in the same amount of light, despite that their aperture sizes are different (they must be different, because they have the same f-number and different focal lengths).

If we use lens A with a certain f-number and lens B with the next f-number in the scale, then during any given time interval image A will gather twice as much light as image B because they differ in just one f-stop.

If we take an image, then increase one f-stop but also double the exposure time (half the shutter speed) and take another image, then both images will have the same exposure. They will have received the same total amount of light: in the first case, double illuminance during half the time, while in the second image is half illuminance but double time. The term exposure value (EV) is used in photography to refer to all pairs of f-number and exposure time values that produce images with the same exposure:

$$EV = log_2\left(\frac{N^2}{t}\right). \tag{2.9}$$

The exposure scale is also discrete, and -1EV or "stop" means doubling the exposure and +1EV halving it. But images with the same exposure will be different, though: the one with the slower shutter will have motion blur and increased depth of field, as Figure 2.26 shows.

FIGURE 2.26: Two images with the same exposure. Left: $N = 2.8$, $t = \frac{1}{50}s$. Right: $N = 8$, $t = \frac{1}{6}s$.

The above derivation of the f-number assumed that all light reaching the lens goes through it. In practice, though, reflection and absorption are unavoidable and it's a common sceneario that only around 80% of the light that enters the camera actually reaches the sensor [235]. F-numbers can be corrected to account for the transparency of the optical elements: these are the T-numbers (T for transparency), also arranged in a scale of stops or T-stops.

2.6.2 Depth of field

If we focus the camera on a certain object O which is at a distance d_O from the camera, what we are doing is moving the lens system so that the image of the object is formed exactly on the image plane, where the sensor is located. Objects at distances larger or smaller than d_O will have their images formed behind or in front of the image plane, therefore they will appear blurry, i.e. points will be imaged as circles, as Figure 2.22 showed. But if these circles are sufficiently small, in particular smaller than the size of a single element cell of the sensor (*"the size of a pixel"* if you will), then the objects will be perceived as sharp, not out of focus. If we move O away, we'll still see it in focus until we reach a certain distance d_f beyond which O will appear blurry: d_f will be the farthest focus distance. If we move O closer to the camera, there'll be a distance d_n such that if O is closer than d_n it will not appear sharp: d_n will be the nearest focus distance. The depth of field (DOF) is the distance between the farthest and the nearest scene objects that appear in focus, i.e. $DOF = d_f - d_n$.

We will now compute the DOF for a camera where the lens has focal distance f and aperture diameter D, and d is the maximum value for the diameter of a blurred circle to be perceived as a point. In this section we are following the derivation in [26].

Figure 2.27 shows a point A at distance O from the lens: its image will also be a point because it's located at distance I, which is the position of the image plane. Points B and C, located at O_1 and O_2 respectively, will have their images at positions I_1 and I_2 which are different from I. This implies that the light rays from point B will cross behind the image plane, and therefore their intersection with the image plane will form a circle, called the circle of confusion. The same happens with point C. The positions of points B and C are chosen so that their circles of confusion both have the same diameter d, which we said was the maximum acceptable diameter value for these circles to appear as points in the image. Hence, points at distances larger than O_2 or smaller than O_1 will appear as blurry circles in the image, and therefore the far focus distance is $d_f = O_2$ and the near focus distance is $d_n = O_1$. To compute the DOF, then, we must compute O_1 and O_2.

The lens-maker's Equation $\frac{1}{O} + \frac{1}{I} = \frac{1}{f}$ for points A, B and C gives us:

$$I = \frac{Of}{O - f}, \tag{2.10}$$

$$O_1 = \frac{I_1 f}{I_1 - f}, \tag{2.11}$$

and

$$O_2 = \frac{I_2 f}{I_2 - f}. \tag{2.12}$$

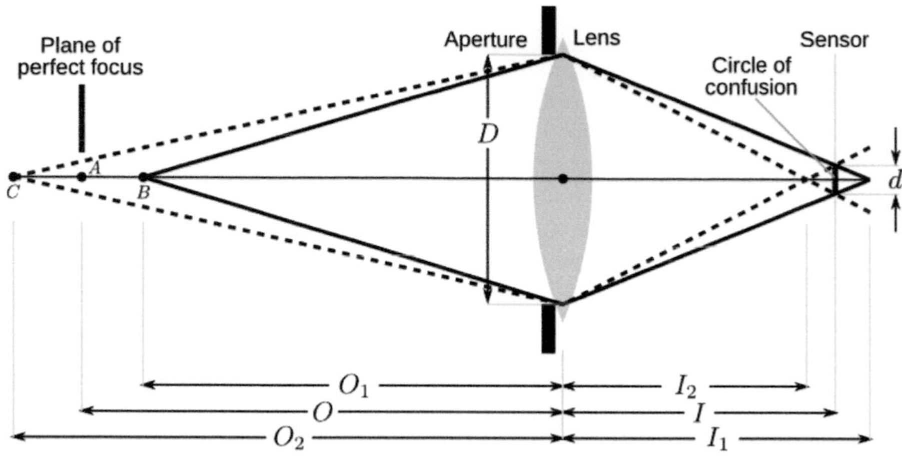

FIGURE 2.27: Depth of Field. Adapted from [26], original figure by Dr. Leslie Wilk.

The transmitted rays from point B cross at I_1, and similar triangles give us:

$$\frac{d}{I_1 - I} = \frac{D}{I_1}, \tag{2.13}$$

therefore we have:

$$I_1 = \frac{DI}{D - d}. \tag{2.14}$$

Plugging I from Equation 2.10 into Equation 2.14, using the resulting value of I_1 in Equation 2.11 and finally using the lens-maker's Equation for A, we get:

$$O_1 = \frac{O}{1 + X}, \tag{2.15}$$

where

$$X = \frac{dO}{DI}. \tag{2.16}$$

The transmitted rays from point C cross at I_2, and similar triangles give us:

$$\frac{d}{I - I_2} = \frac{D}{I_2}, \tag{2.17}$$

therefore we have:

$$I_2 = \frac{DI}{D+d}. \tag{2.18}$$

Plugging I from Equation 2.10 into Equation 2.18, using the resulting value of I_2 in Equation 2.12 and finally using the lens-maker's Equation for A, we get:

$$O_2 = \frac{O}{1-X}, \tag{2.19}$$

where X is the same as for O_1.

We can now compute the DOF:

$$DOF = O_2 - O_1 = \frac{2OX}{1-X^2}, \tag{2.20}$$

where we can introduce the f-number $N = \frac{f}{D}$, the magnification $M = \frac{I}{O} = \frac{f}{O-f}$ and use the approximation $\left(\frac{d}{D}\right)^2 \simeq= 0$ to obtain the DOF equation:

$$DOF = 2dN\frac{M+1}{M^2}. \tag{2.21}$$

There are several important properties for photography that we can learn from this equation:

- At a fixed focal length (and therefore fixed magnification), increasing the f-number (reducing the aperture) increases the depth of field, as Figure 2.28 shows.

- At a fixed f-number, increasing the magnification (by increasing the focal length, i.e. zooming-in) makes the depth of field decrease, as Figure 2.29 shows.

- If we fix both the relative aperture (f-number) and object size (magnification), the DOF remains constant regardless of the focal length, as Figure 2.30 shows. That is, if we shoot a subject and then move back and increase the focal length so that the subject remains the same size, also keeping the f-number constant, then the DOF will be the same as before. This may come as a surprise, because there is the extended myth among many photographers that the DOF decreases with increasing focal length.

- From Equation 2.19 we see that when $X = 1$ then $O_2 = \infty$: if the far focus distance is infinite, then everything behind the object will be sharp, in focus. The condition $X = 1$ holds when the object is located at $O = H$, where H is called the *hyperfocal distance*:

$$H = \frac{f^2}{Nd}\left(1 + \frac{d}{D}\right). \tag{2.22}$$

If $O = H$ then $X = 1$, $d_f = O_2 - O = \infty$ and $d_n = O - O_1 = \frac{H}{2}$, therefore if the camera is focused on an object placed at the hyperfocal distance from the camera, everything from $\frac{H}{2}$ to infinity will be in focus [26]. For instance, for a large-sensor digital camera with $d = 0.03mm$, focal length $f = 32mm$, f-number f/5.6, and disregarding $\frac{d}{D} \simeq 0$, we have $H = 6.4m$

FIGURE 2.28: Effect of f-number on DOF. Left: f/2.8. Right: f/8.

FIGURE 2.29: Effect of magnification on DOF.

FIGURE 2.30: Focal length changes but DOF remains constant if f-number and magnification are constant.

2.6.3 Prime vs. zoom lenses

Lenses can have fixed focal length (prime lenses) or the focal length may vary (zoom lenses). With prime lenses, the only way to modify the size of an object in the image is to move the camera. A zoom lens, on the other hand, allows us to change the focal length of the system (and hence the magnification and angle of view) while keeping the camera in the same position. This can be achieved because zoom lenses are actually comprised of several prime lenses which, by moving axially in a certain way, allow the focal length to vary and keep the plane of focus fixed (zoom lenses in which the plane of focus changes with the focal length are called *varifocal lenses*).

In its simplest configuration, a zoom lens consists of an afocal system followed by a prime lens. When a collimated beam (parallel lightrays) comes to the afocal system along the optical axis, its direction is not altered but its width is; see Figure 2.31. Since the beam direction is not modified, the afocal system does not focus light and hence its name.

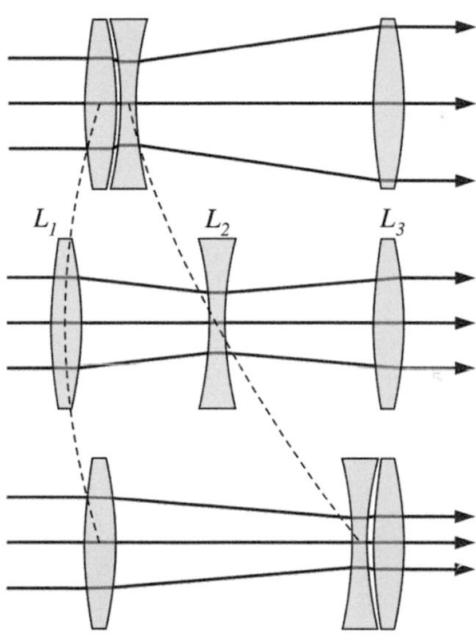

FIGURE 2.31: Afocal system. Figure from [8].

But the afocal system is effectively changing the size of the objects by modifying the size of the beam. Next, the altered beam is focused by a converging, prime lens; see Figure 2.32.

Recall that image magnification can be computed as $M = \frac{f}{O-f}$, where O is the object distance, and because normally $O >> f$ we get that increasing the focal length by a given factor we are also increasing the image size by the same factor. This is why zoom lenses are commonly described by the ratio of the longest to the shortest focal length the lens sustains, and this ratio is also the zoom factor: for example, a zoom lens with focal lengths ranging from 100 mm to 400 mm may be described as a 4:1 or "4" zoom. Because of the technical challenges in correcting optical aberrations for the entire range

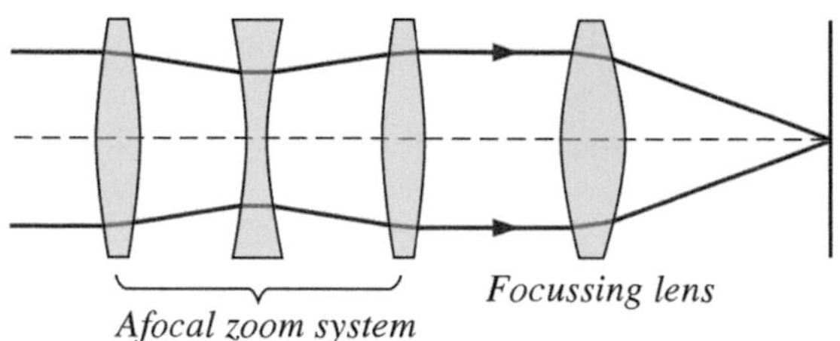

FIGURE 2.32: Diagram of a simple zoom lens. Figure from [21].

of focal lengths the zoom lens works with, zoom lenses with ratios above 3x produce images of lesser quality than prime lenses. This is mainly an issue in still photography because image defects are less visible with moving images and lower resolutions, as in TV [21].

2.6.4 Modulation transfer function

The modulation transfer function (MTF) of a lens measures its capability to transfer detail from object to image. Transfer functions characterize systems that produce an output given an input. In an optical system, detail is given by the resolution, usually specified in line-pairs (alternating light and dark stripes) per millimeter, LP/mm [78]. Because of the optical aberrations mentioned earlier, as details become finer the image that the lens produces becomes less sharp; see Figure 2.33.

FIGURE 2.33: Left: object. Right: image produced by lens. Images from [33].

Sharpness can be measured through contrast, which is defined as

$$Contrast = \frac{I_M - I_m}{I_M + I_m}, \tag{2.23}$$

where I_M and I_m stand for the brightest and darkest image values, respectively. The left image in Figure 2.33 has a contrast of 1 or 100%, but the image on the right has a contrast lower than 1 and decreasing with line-pair width, because as the black and white stripes become thinner their images have a lower maximum value and a higher minimum value. MTF is a plot of contrast, measured in percent, against spatial frequency measured in LP/mm

[78]. The highest line frequency that a lens can reproduce without losing more than 50% of the contrast is called the "MTF-50" value and it correlates well with our perception of sharpness [33].

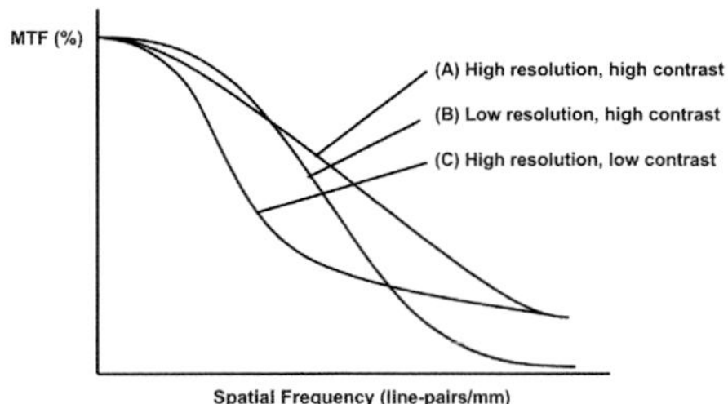

FIGURE 2.34: MTF plots for different types of lenses. Image from [235].

Figure 2.34, taken from [235], plots the MTF for three different types of lenses A, B and C. Lens A has high MTF values for all frequencies and therefore its images will be good in terms of contrast and resolution. Lens B has an MTF that decays abruptly, producing good contrast for low and medium frequencies but loss of resolution for high frequencies. Lens C has an MTF that decays slowly, and while this is good in terms of resolution it also implies that contrast will be very similar for medium and high frequencies, i.e. the overall contrast will be low, with no modulation between light and dark [235].

In practice, though, the MTF is not expressed as contrast vs. spatial frequency. Commonly MTF is measured along a line from the image center to an image corner, for a fixed line frequency such as 10LP/mm [33]. See Figure 2.35. Because of the optical aberrations described before, the MTF will always be higher at the center. Also, the MTF curve gets more flat as we increase the f-number, precisely because several aberrations are corrected by decreasing the aperture: compare in Figure 2.35 the solid bold blue line (f/8) with the solid bold black (aperture wide open).

This last result might give us the impression that the MTF can be optimized just by increasing the f-number at will. But this is not the case, as Figure 2.36 shows: the MTF-50 does indeed increase with f-number, but starting from f/11 it starts to decrease. This is due to the effects of diffraction, because at the scales of extremely small apertures the wave nature of light can no longer be ignored and with wave optics the image of a point is no longer a point but a disc, whose radius depends on the f-number. Furthermore, in current digital cameras the pixel pitch is only five times the wavelength of light, and diffraction effects are then unavoidable [235].

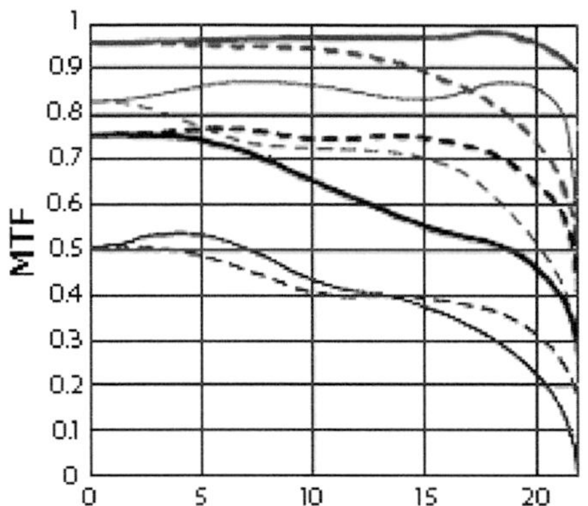

FIGURE 2.35: MTF plots for a Canon 35mm f/1.4 prime lens. Image from [33].

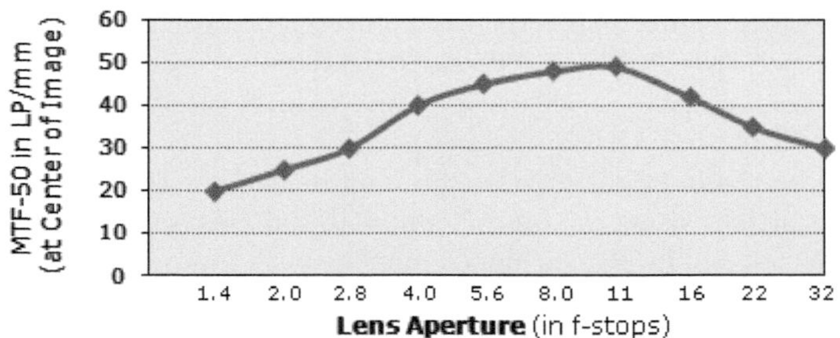

FIGURE 2.36: MTF-50 in LP/mm as a function of f-number. Image from [33].

Part II

Camera

Chapter 3

Camera

3.1 Image processing pipeline

The image processing pipeline of a camera is the sequence of steps in which a digital image is formed, and both the ordering as well as the particular algorithm used in each stage are important [320, 223]. Though camera makers usually don't make public this information, digital cameras commonly perform the following operations, which we'll introduce in this chapter and then discuss in more detail in the remainder of the book: demosaicking, black level adjustment, white balance, color correction, gamma correction, noise reduction, contrast and color enhancement, and compression. At the end of the chapter we'll briefly discuss criteria for the ordering of these operations. Figure 3.1, taken from the survey by Ramanath et al. [320] shows the schematic of a pipeline.

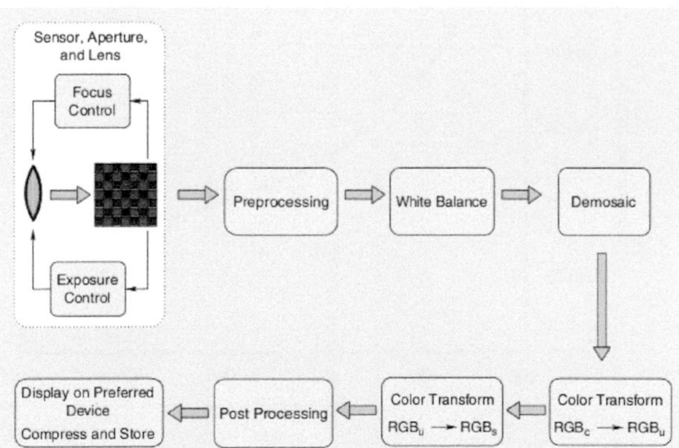

FIGURE 3.1: Image processing pipeline. Figure from [320].

3.2 Image sensors

The first stage is that of image acquisition, performed by the camera sensor(s). An image sensor is a semiconductor device that uses the photoelectric effect to convert photons into electrical signals [291]. It is formed by an array of cells (picture elements or pixels); at each of these locations, incident photons raise energy levels in the semiconductor material, letting loose electrons and creating electrical charge [358]. The relationship between number of photogenerated electrons and number of photons is linear, but the subsequent electronic processes (that convert electrons into amplified output voltage) may introduce nonlinearities. The proportion of photons absorbed decays with the wavelength of the incident light, and with wavelengths longer than $1100nm$ silicon behaves as a transparent material, as Figure 3.2 (from [291]) shows. But since the visible spectrum ends at $700nm$, it is necessary to put in the optical path an infrared (IR) filter that prevents wavelengths higher than $700nm$ to reach the sensor. This can be achieved by coating an optical surface (the lens or some other optical component in the optical path) with layers of dielectric materials that reflect IR light, in what is known as a *hot mirror*; alternatively, the filter could absorb IR radiation, in which case there's the potential problem of overheating [67].

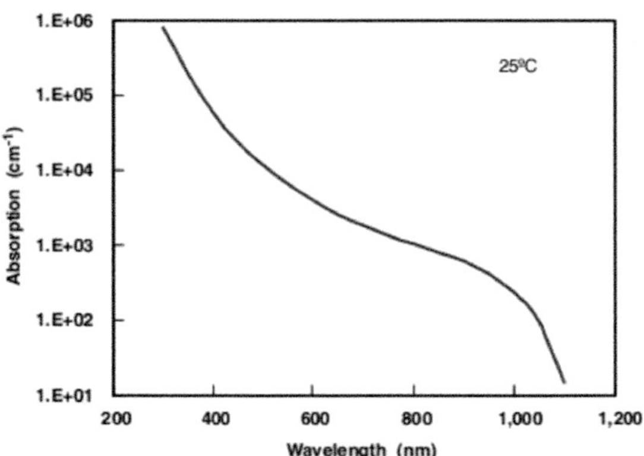

FIGURE 3.2: Absorption coefficient of light in silicon. Figure from [291].

3.2.1 Pixel classes

Pixels can be of two different classes:

- Photodiodes, which store the electrical charge around metal junctions created by implanting ions into the silicon.
- Photogates, which store the electrical charge in "potential wells" created by capacitors.

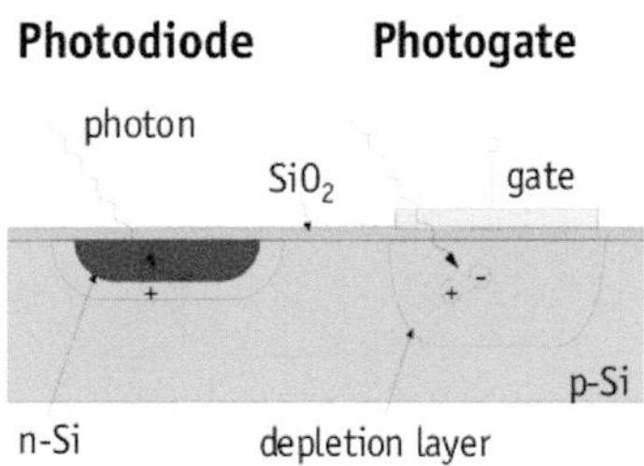

FIGURE 3.3: Types of picture elements. Figure from [358].

A most important characteristic in a pixel is its fill factor, which is the portion of the pixel area that is photosensitive. The major advantage of photogates is their very high fill factor, of almost 100%, allowing them to convert more photons and generate a larger signal. As a downside, the polysilicon gate over the pixel reduces its sensitivity in the blue end of the spectrum [358]. Photodiodes, on the other hand, have better sensitivity but they are more complex and require the presence of opaque circuit elements (transfer gate, channel stop region to isolate pixels, shift register to move the charge) that considerably reduce the fill factor to 30-50% [291].

In order to overcome these limitations, photogates may use very thin polysilicon gates so as not to compromise sensitivity, and photodiodes may use microlenses to increase fill factor. A microlens array is placed on top of the photodiode array, and each microlens focuses incident light onto the photosensitive part of the pixel, effectively increasing fill factor to around 70%; see Figure 3.4. This approach noticeably enhances sensitivity, but it's not free of shortcomings. Its main disadvantage is that it produces shading in the image, because now the percentage of photons that are focused onto the photosensitive part of the pixel depends on the angle of incidence of the light; see Figure 3.5. Possible solutions to this issue include decreasing the distance between the microlens array and pixel surface, shifting the position of the microlenses depending on image location, and adding another layer of microlenses [291]. Furthermore, with large apertures the microlens array may cause vignetting, pixel crosstalk, light scattering and diffraction problems [208]. Some of these problems can be reduced by means of image processing after capture [358].

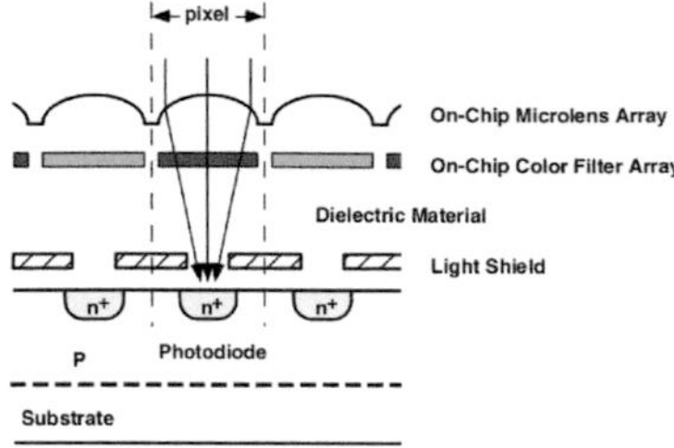

FIGURE 3.4: Scheme of photodiode with microlens array. Figure from [291].

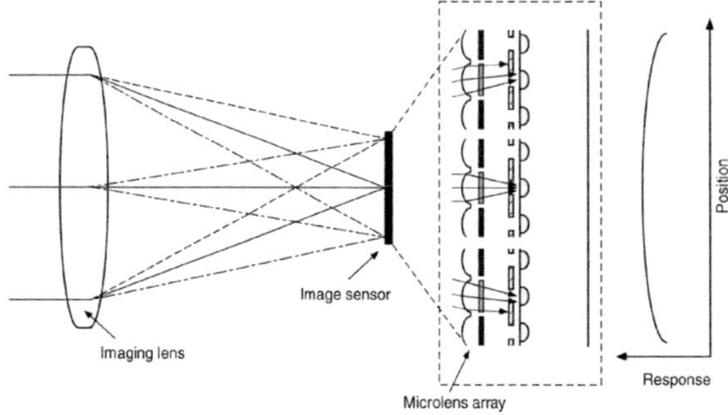

FIGURE 3.5: Shading caused by microlens array. Figure from [291].

3.2.2 Sensor classes

The electrical charge is accumulated while the sensor is being exposed to light, and then it must be converted into voltage or current through the scanning of the image array. There are two large families of sensor devices (both using photodiodes and photogates) and their differences arise from the way they perform this scanning:

- A Charged Coupled Device (CCD) transfers the charge vertically from pixel to pixel over the entire array, then it is transferred horizontally and converted into voltage at just one output amplifier. See Figure 3.6, left.

- CMOS (Complementary Metal Oxide Semiconductor) devices perform the charge-to-voltage conversion at each pixel location. See Figure 3.6, right.

These two approaches imply very different implementation strategies, as we will discuss.

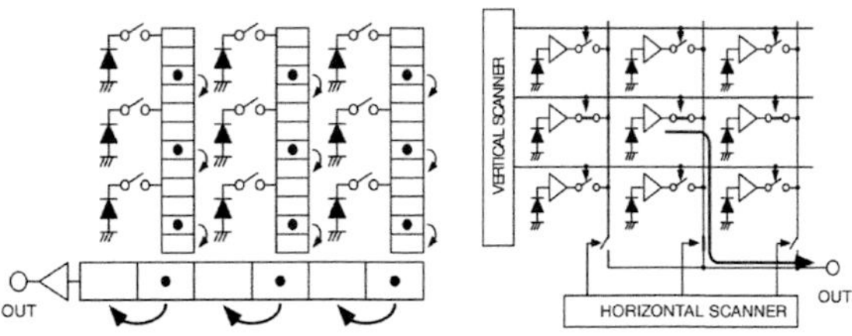

FIGURE 3.6: Scanning the image array. Left: CCD transfer scheme. Right: CMOS X-Y address scheme. Figure from [291].

CCDs have a very high uniformity in their output, because they use a limited number of amplifiers; on the other hand, these amplifiers must be able to work at the very high bandwidth of cinema (several millions of pixels per second) under the limitation of amplifier noise. CMOS devices have at least one amplifier per pixel, which has advantages (these amplifiers can be simple, with low bandwidth and low power consumption, allowing for higher frame rates) and inconveniences (it's difficult to make milllions of amplifiers in a uniform way, so CMOS devices suffer from higher fixed-pattern noise; amplifiers reduce the fill-factor, therefore microlens arrays are almost always used with CMOS) [358].

Most sensors output analog signals that are later digitized at a camera module. Given that in principle CMOS sensors are smaller and have lower consumption than their CCD counterparts, the analog-to-digital conversion can be implemented at the sensor along with some image processing functionality (the so-called "camera on a chip" approach) although this causes complexity and optimization problems which explains why most successful CMOS imagers don't perform analog to digital conversion on chip [358].

3.2.3 Interlaced vs. progressive scanning

There are two modes of scanning: progressive and interlaced. In the progressive mode all lines in an image are read in sequential order, one after the other. In the interlaced mode the odd lines are read first, and then the even lines, thus creating two image *fields* for every image *frame*. Interlaced scanning has its origins in the early days of television, when it was devised as a way of achieving an acceptable compromise between frame-rate, image resolution and bandwidth requirements. It was incorporated into all major broadcasting standards, and because of backward-compatibility policies it still is very much in use, although definitely not in professional cinema. But for non-professional cinema as well as for legacy material there is the need of performing interlaced to progressive conversion, since current display technology (unlike, say, CRT TV sets) produces a progressive output. We'll discuss this in Chapter 6.

Figure 3.7 shows examples of the same image acquired under these different modes. Notice how motion is quite problematic both for interlaced and raster progressive scanning. This latter case is the rolling-shutter problem of simple CMOS imagers, which we'll comment on later.

 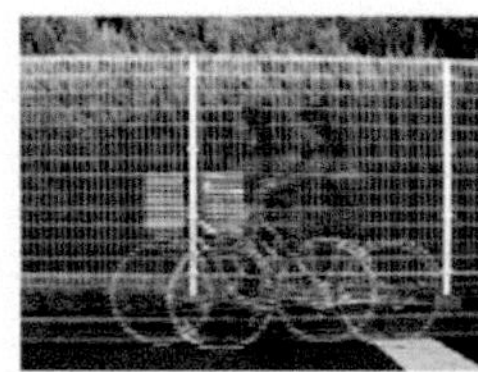 

FIGURE 3.7: Left: progressive scanning, raster acquisition (line by line). Middle: interlaced scanning. Right: progressive scanning, simultaneous acquisition. Figure from [292].

But for image sensors, interlaced scanning has the advantage of allowing us to average line values and hence increase sensitivity (at the expense of reducing vertical resolution) [387]. Also, in CCDs interlaced scanning requires simpler circuitry, permitting a larger resolution [292].

3.2.4 CCD types

There are three main types of layouts for CCDs [358]:

- Full frame CCDs use photogates and have the simplest sensor layout with the highest fill factor. A mechanical shutter is required to block the light during charge-transfer (otherwise the image would be smeared while the sensor is reading it line by line) but this is not a problem in cinematography since most cameras come equipped with a rotating shutter.

- Frame transfer CCDs also use photogates. They have a duplicated frame, the regular full frame plus a storage region to which the image is transferred (at high speed) and where it is read out, while at the same time the next image is being acquired. This layout improves on speed and reduces smear, at the price of duplicating silicon and increasing power dissipation.

- Interline transfer (ILT) CCDs use photodiodes which, as we mentioned, provide good sensitivity for the blue part of the spectrum at the price of reducing fill factor due to the opaque circuit elements (these elements allow for electronic shuttering, but in digital cinema this is not really an improvement since cameras already come with mechanical shutters), which is why most ILT CCDs use microlens arrays.

The charge transfer scanning procedure of CCDs makes them subject to blooming, which occurs when charge overflows a pixel (e.g. if there is a bright light source in the shot) and spills vertically; see Figure 3.8. In order to perform antiblooming, an overflow drain must be added.

FIGURE 3.8: Left: blooming. Right: CCD with overflow drain. Figure from [292].

3.2.5 CMOS types

In their most simple configuration, CMOS sensors have one transistor per (photodiode) pixel. This "passive pixels" architecture provides a good fill factor (though still much lower than with CCDs) at the price of a low signal to noise ratio. To reduce the noise, an amplifier is added to each pixel in the "active pixels" layout: as each amplifier is made with three transistors, this configuration is also referred to as 3T. Noise response is better than with passive pixels, but also lacking. To improve it, the "pinned photodiode" configuration requires an extra diode (so it's also called 4T) but allows us to perform Correlated Double Sampling (CDS) [130]. CDS is a noise reduction technique consisting of sampling two images, one with the shutter closed and another after exposure, and subtracting the latter from the former, thus reducing *dark current* noise (we'll discuss noise sources below). An extra transistor (5T) allows us to perform global shuttering, i.e. to cease exposure on all pixels simultaneously, thus avoiding image artefacts related to fast motion [358]. Without global shuttering, images with motion are distorted by the so-called *rolling shutter effect*: rows are read sequentially, and the time lag between rows may cause noticeable visual artifacts such as stretching of objects when motion is horizontal; see Figure 3.9.

Transistors are opaque, therefore most CMOS devices use microlens arrays. Microlenses with CMOS sensors pose worse problems than with CCDs: on top of the silicon surface is a stack of transistors and circuitry, which forces the microlens to be further away from the pixel thus compromising color response and introducing a strong dependence of sensitivity on the angle of incidence

FIGURE 3.9: The rolling shutter effect. Left: image with static camera. Right: image when camera moves.

of the light. Also, more involved circuitry requires piling up more layers on top of the substrate which increases noise and optical cross-talk [130].

3.2.6 Noise in image sensors

The amount of noise present in the image signal is a very important characteristic of the sensor device. The dynamic range of the sensor is a magnitude that conveys its ability to capture detail both in bright and dark regions of the image simultaneously, and it is described using the ratio between the pixel's largest charge capacity and the noise floor; it is expressed as a ratio (e.g. 4096:1), in decibels, or bits [358] (we discuss dynamic range in Chapter 11). Noise can be classified as constant in time (fixed-pattern noise or FPN) or varying in time (temporal noise) [291].

FPN is caused mainly by dark current and shading, and being constant it can not be eliminated by Correlated Double Sampling (CDS). Dark current is a parasite current, not originated from photon conversion, that nevertheless is integrated as charge. This current may be originated inside a pixel, or in a transfer channel of a CCD sensor, or at an amplifier in a CMOS device. It increases with exposure time and with temperature, reducing dynamic range (by increasing the output level corresponding to "dark"). The borders of image arrays are composed of "optical black pixels," which are never exposed so that they can be used to estimate dark current levels and therefore a proper black level for the image. Shading is a slow spatial variation in the brightness of the image, which may be originated by a local heat source (which in turn generates dark current), a microlens array (as discussed above), non-uniformity of electrical pulses in CCDs, or nonuniformity of bias and ground in CMOS sensors [291].

Temporal noise manifests in three kinds of noise: thermal, shot and flicker, both in CCD and CMOS sensors. Thermal noise is due to thermal motion of electrons inside a resistance. MOS transistors behave as a resistance during exposure, making thermal noise appear (this is called "reset noise" or "kTC noise"); readout electronics are also a thermal noise source [291]. Shot noise is caused by the discrete nature of incident photons and generated electrons. The arrival of photons follows a Poisson distribution which, if the number of particles is not high, exhibits fluctuations that translate into variations in brightness called photon shot noise. Dark current shot noise is the same type of phenomenon, in this case associated to the random generation of charge in the pixel. Flicker noise, also called 1/f noise or pink noise, is originated by the properties of the materials that make up amplifiers, and exhibits a spectral distribution that is inversely proportional to frequency (hence the 1/f name) as opposed to thermal and shot noise, which have a uniform spectral distribution (noise with a uniform spectral distribution is called "white" in analogy with white light, whereas 1/f noise is called "pink" because that's the appearence of light with a 1/f distribution).

We'll discuss denoising algorithms in Chapter 5.

3.2.7 Capturing colors

As we have seen, image sensors work by transforming photons into electrical signals, but this is only a measure of the incident light intensity, not of its color. Photons don't carry color information, since there is only one kind of photon; color is a property of the wavelengths of the light. In order to capture colors, two main configurations are in use [67]:

- Three-sensor systems. Incoming light is separated into three different color channels using a beam splitter (a type of prism) and there is one sensor devoted to each. See Figure 3.10 (left).
- Color Filter Arrays (CFA). In this one-sensor approach, the image array is covered by a mosaic color filter of three colors, making each pixel in the array capture one color channel only. The CFA is added to the sensor either by a pigment process or a dye process (pigment-based CFAs are more resistant to heat and light). See Figure 3.10 (right).

Practical problems abound with the three-sensor approach: sensors must be very carefully aligned to avoid chromatic aberration, the simple refraction of beamsplitters may not be enough for precise color separation, and the optical path through the prism increases both lateral and longitudinal aberration [358].

As for CFAs, color choices for the filter are primary colors (red, green and blue) and complementary colors (cyan, yellow and magenta); RGB CFAs are superior in terms of color separation, color reproduction and signal to noise ratio. By far the most popular CFA is the Bayer pattern, using an RGB configuration with two green pixels per red or blue pixel (with the rationale that the human visual system is more responsive to light of medium wavelengths) [291]. The downside is that, since for each pixel we only have color information for one channel, an interpolation process called demosaicking must be carried

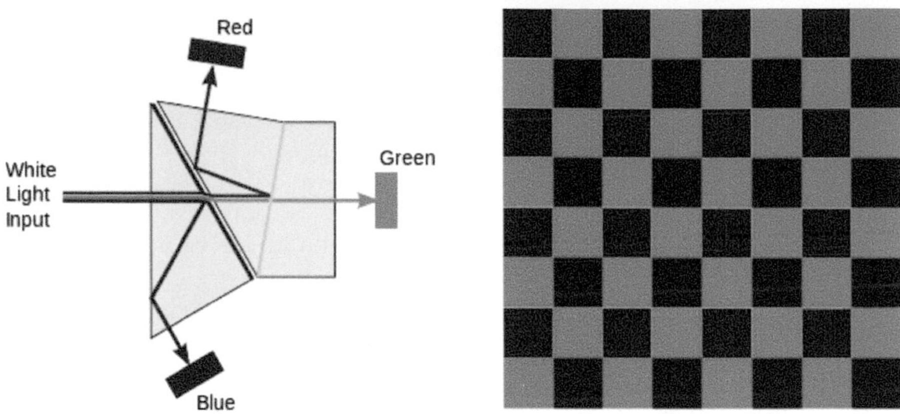

FIGURE 3.10: Capturing color information. Left: a three sensor system. Right: Bayer CFA over a single sensor. Left image adapted from [15].

out to estimate the values of the other two channels. Furthermore, in the presence of aliasing the demosaicking algorithms may cause substantial visual artifacts, especially if the aliasing is not taking place in the three channels. Demosaicking is discussed in Chapter 6.

In order to avoid this, antialiasing filters are used, mainly based on one of these two techniques [67] illustrated in Figure 3.11:

- polarization: some pieces of birefringent material split an incoming light ray into several beams, each going to a different pixel;
- phase delay: the surface of an optical element is etched with a pattern, so that some light goes through more filter material, suffering a phase delay and interfering with the light going through less filter material, thus reducing higher frequencies.

These antialiasing elements are called optical low-pass filters (OLPF), and they provide the only effective way of reducing aliasing, which can't be handled adequately in post-production [64]. The use of OLPFs is essential in movies shot with DSLR cameras, where the sensor may have 10-20 megapixels from which the 2 megapixels HD video images are obtained simply by decimation, i.e. just by downsampling, not by averaging pixels with their neighbors [143]. Therefore, unless an OLPF is used, the final HD video is practically guaranteed to show aliasing.

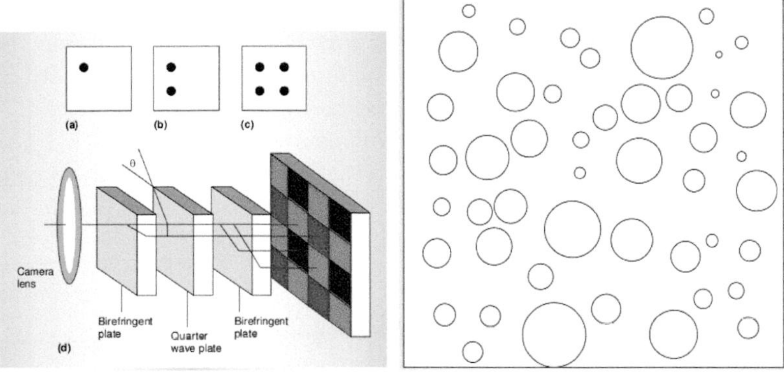

FIGURE 3.11: Left: antialiasing based on polarization, (a) to (c) are the split beams, (d) shows the full light path. Right: phase delay antialiasing filter. Images from [67].

3.3 Exposure control

Exposure is the amount of light that is allowed to reach the sensor while capturing an image. In order to regulate it we may modify the aperture (varying the f-number) and/or the shutter speed (which defines the exposure time). In photography, an *exposure value* or EV denotes all combinations of relative aperture and shutter speed that provide the same exposure:

$$EV = log_2\left(\frac{F^2}{t}\right), \tag{3.1}$$

where F is the f-number and t is the exposure time (in seconds), and the quantity $\frac{F^2}{t}$ is proportional to the exposure. For instance, if we double the exposure time but divide by two the aperture area, the exposure remains the same and so does the EV. If we just double the exposure time while keeping the relative aperture fixed, then the exposure is also doubled and the EV decreases in one unit (one "stop"). The optimum exposure for a given image is clearly image-dependent, e.g. more brightly lit scenes will require a larger EV.

Automatic exposure control is the set of processes by which the camera modifies aperture and shutter speed (and sometimes sensor gain) in order to expose each image frame correctly, i.e. avoiding overexposure of bright image regions and underexposure of dark regions. Cinema professionals prefer manual exposure control, mainly because automatic controls do not allow the cinematographer to choose freely the aesthetics of the image, but also because the automatic exposure may degrade the quality of the image, increasing noise when compensating for low light levels, modifying the depth of field when changing the aperture, or creating annoying oscillations in the average luminance in highly dynamic scenes [143].

As explained in [88], most auto-exposure algorithms work by taking a (temporary, not to be recorded) picture with a pre-determined exposure value EV_{pre}, computing a single brightness value B_{pre} from that picture, selecting an optimal brightness value B_{opt}, computing the optimal exposure value thus:

$$EV_{opt} = EV_{pre} + log_2\left(\frac{B_{pre}}{B_{opt}}\right), \tag{3.2}$$

and finally modifying aperture, shutter speed and gain so as to take an exposure-corrected picture with exposure value EV_{opt}.

3.3.1 Exposure metering

There are several strategies for metering [135]:

- Zone metering. A single output measurement is obtained as the weighted average of contributions from a number of zones, with higher weights for central areas. The most common approach is to compute the average luminance; but median, mode (of the histogram distribution of the luminance) or peak white measurements (to avoid clipping of very bright areas) can also be used.

- Matrix or multi-zone metering. Light intensity is measured in several points in the scene, and several factors are taken into account to derive the best exposure value: autofocus point, distance to subject, areas in focus or out of focus, colors, backlighting, and, very importantly, matchings with a pre-stored database of thousands of exposures. This metering system is mainly used with still cameras, as its complexity may make it unstable with dynamic scenes.

- Content-based metering. These systems try to expose optimally those regions in the image that are deemed most relevant, using measures such as contrast, focus, skin-tones, object-based detection and tracking, etc.

3.3.2 Control mechanisms

As mentioned above, there are three ways to control the exposure, by modifying the aperture, shutter speed and sensor gain.

The aperture control may be performed entirely at the lens (AC iris) with an integrated amplifier changing the iris area in order to keep constant the exposure; or the signal may come from outside the lens unit (DC iris). The latter is the preferred approach for high-end video; difficulties arise from the fact that the camera must provide signal and control mechanisms that should work with any DC iris lens. Modifying shutter speed and sensor gain provides stable and fast (next frame) exposure control; also, reducing exposure time decreases motion blur, which is something that can't be achieved via aperture control and is critical in rolling-shutter CMOS sensors. On the other hand, while integration time may change by a factor of 1000 at the most, apertures have a much wider range, and if the amount of light is large it is better to reduce the aperture so that less light reaches the sensor, thus avoiding deterioration of the dye materials and burn-in effect. But each option has its problems: short time exposures cause flicker, and small apertures increase depth of field and may not be too small in order to avoid diffraction problems. The output signal level is regulated also by the sensor gain. While increasing exposure time increases SNR, increasing the gain does not change SNR but it does increase the amplitude of the noise. Therefore, it is often preferred to leave gain control as a last option if controling exposure time and aperture is not sufficient [135].

3.3.3 Extension of dynamic range

Scenes with a high dynamic range, with details both in bright and dark regions, may require a sensor with a dynamic range of 100dB, whereas a good CCD or CMOS sensor may be able to provide 74dB. As we mentioned above, reducing the noise floor is essential to increase the dynamic range of the signal in standard CCD and CMOS sensors. There are other alternatives, like using non-linear (log-shaped) response sensors, dual-pixel sensors (each pixel is split into two parts of different responsiveness), or exposure bracketing (long and short exposure frames are captured and then merged). But none of them really works, having color fidelity, sensitivity or ghosting problems, so HDR video remains an open problem [135].

3.4 Focus control

Automatic focus control (AF) is the process by which the distance between sensor and optical system is regulated, so that the image of the object of interest in the scene is formed on a plane that coincides with the surface of the sensor. Digital cameras may use different AF mechanisms (see [209] and references therein):

- infrared AF, where an infrared ray is used to estimate the distance to the object of interest;
- through-the-lens AF, where the sensor-lens distance is adapted until the signals from both the upper and the lower part of the lens are in phase [320];

– contrast detection AF, where the degree of focus is estimated as the amount of high-frequency information in the image (since blurry regions have lower frequency components).

Nevertheless, and as was the case with automatic exposure, cinema professionals absolutely avoid the autofocus: deciding what is in focus and what is not is a crucial artistic choice, and camera operators usually have an assistant (focus puller) who, as the shot progresses, manually changes the focus making sure it evolves in accordance with the director's intentions [143].

3.5 White balance

We've discussed earlier in Chapter 1 how an essential property of human vision is our ability to perceive as constant the color of an object despite variations in the illumination conditions. We see a white shirt as being the same white under daylight as under (yellowish) tungsten light or under (bluish) fluorescent light. The same happens if the color of the shirt is other than white; we don't see it changing under changing illuminants. But the light coming from the shirt and reaching our eyes does indeed change with the illumination, that is, different *sensations* (e.g. white, yellowish and bluish light) produce the same *perception* of color (white). Furthermore, in the same scene we may have two objects producing the same sensation (the light coming from them has the same intensity and spectral power distribution) but we perceive them as being of different colors. It is clear then that our visual system is doing more than measuring the light at each location in the image; therefore we need our cameras to mimic this behavior; otherwise, if we just used the triplets of colors as captured by the sensor, the colors would never appear right except under the exact lighting conditions used during the callibration of the camera [67].

This problem is usually considered with several simplifying assumptions: all objects in the scene are flat, matte, Lambertian surfaces (diffuse surfaces, which reflect light evenly in all directions) and uniformly illuminated [170], with a single illuminant. In this scenario, the triplet of RGB values captured at any given location of the sensor is:

$$\begin{aligned} R &= \int_{380}^{780} r(\lambda)I(\lambda)S(\lambda)d\lambda \\ G &= \int_{380}^{780} g(\lambda)I(\lambda)S(\lambda)d\lambda \\ B &= \int_{380}^{780} b(\lambda)I(\lambda)S(\lambda)d\lambda, \end{aligned} \tag{3.3}$$

where $r(\lambda), g(\lambda)$ and $b(\lambda)$ are the spectral sensitivities of the red, green and blue filters used by the camera, $I(\lambda)$ is the power distribution of the illuminant, and $S(\lambda)$ is the spectral reflectance of the object.

Experiments in color constancy [214] indicate that our perception of the color of an object is in many cases independent of the illuminant and matches pretty well the reflectance values of the object. Therefore, if the camera is to replicate this behavior, it must "discount the illuminant" from the observed RGB values. Another assumption used in practice is to treat the spectral sensitivity functions $r(\lambda), g(\lambda)$ and $b(\lambda)$ as if they were delta functions centered at the peak sensitivities λ_R, λ_G and λ_B, respectively. This is *not* an accurate representation of these functions, which are very much spread over the visible spectrum as we see in Figure 3.12, but it is a useful simplification for the problem at hand, good enough for many contexts [167], which we will discuss in more detail in Chapter 7. Under this hypothesis then, the equations 3.3 become:

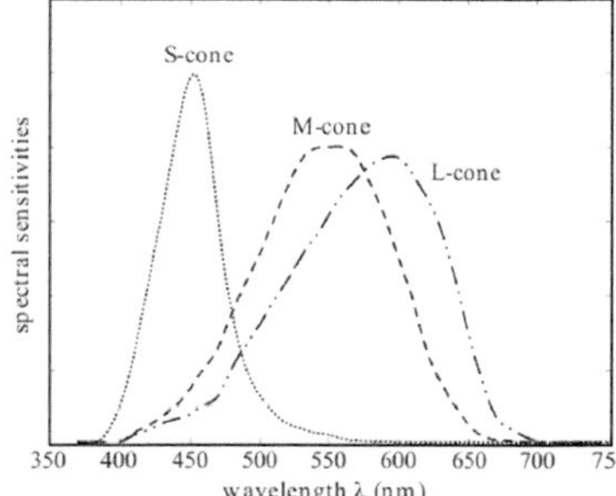

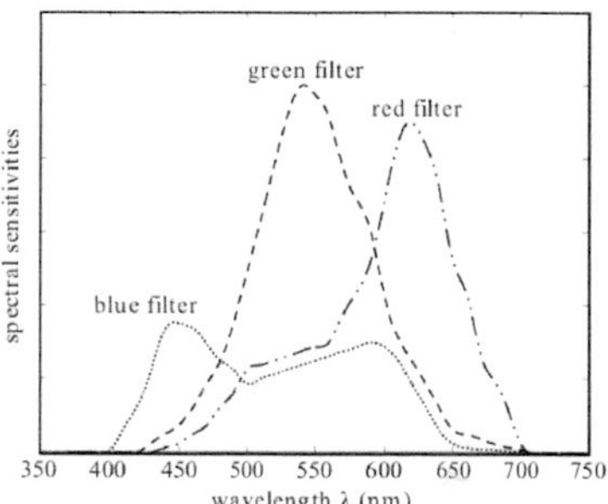

FIGURE 3.12: Spectral sensitivities of: (a) the three types of cones in a human eye, and (b) a typical digital camera. Images from [239].

$$
\begin{aligned}
R &= I(\lambda_R)S(\lambda_R) \\
G &= I(\lambda_G)S(\lambda_G) \\
B &= I(\lambda_B)S(\lambda_B),
\end{aligned}
\tag{3.4}
$$

or, in matrix form:

$$
\begin{bmatrix} R \\ G \\ B \end{bmatrix} = \begin{bmatrix} I(\lambda_R) & 0 & 0 \\ 0 & I(\lambda_G) & 0 \\ 0 & 0 & I(\lambda_B) \end{bmatrix} \begin{bmatrix} S(\lambda_R) \\ S(\lambda_G) \\ S(\lambda_B) \end{bmatrix}. \tag{3.5}
$$

Since the camera observes the values (R, G, B) but we want to have $(S(\lambda_R), S(\lambda_G), S(\lambda_B))$, we need to estimate the illuminant $(I(\lambda_R), I(\lambda_G), I(\lambda_B))$; once we have the illuminant, the reflectances could be recovered with a simple division:

$$S(\lambda_R) = \frac{R}{I(\lambda_R)}, \; S(\lambda_G) = \frac{G}{I(\lambda_G)}, \; S(\lambda_B) = \frac{B}{I(\lambda_B)}. \tag{3.6}$$

The white balance process by which a color corrected triplet (R', G', B') is obtained from (R, G, B) can be written in matrix form thus:

$$\begin{bmatrix} R' \\ G' \\ B' \end{bmatrix} = \begin{bmatrix} \frac{1}{I(\lambda_R)} & 0 & 0 \\ 0 & \frac{1}{I(\lambda_G)} & 0 \\ 0 & 0 & \frac{1}{I(\lambda_B)} \end{bmatrix} \begin{bmatrix} R \\ G \\ B \end{bmatrix}. \tag{3.7}$$

Some cameras proceed this way although it is usually preferred to apply the diagonal model of Equation 3.5 not directly on the RGB values but on their correspondent cone tristimulus values, since color adaptation in the eye is very much dependent on the sensitivity of the cones [205]. We can convert RGB values into CIE tristimulus values XYZ with:

$$\begin{bmatrix} X \\ Y \\ Z \end{bmatrix} = A \begin{bmatrix} R \\ G \\ B \end{bmatrix}, \tag{3.8}$$

where $A = [A_{ij}]$ is a $3x3$ matrix that is chosen for each camera so as to optimize the color reproduction of a given (small) set of colors considered important [205], under a single illuminant, and applied as it is for all the illuminants that can occur [103] (see next section for more details). The color correction procedure can then be written as:

$$\begin{bmatrix} R' \\ G' \\ B' \end{bmatrix} = A^{-1} \begin{bmatrix} \frac{1}{I_X} & 0 & 0 \\ 0 & \frac{1}{I_Y} & 0 \\ 0 & 0 & \frac{1}{I_Z} \end{bmatrix} A \begin{bmatrix} R \\ G \\ B \end{bmatrix}. \tag{3.9}$$

The white balance process consists then of two steps, illumination estimation and color correction. Ideally it should be performed not on-camera but afterwards, as offline post-processing: the rationale, as with demosaicking and denoising algorithms, is that with offline postprocessing we have much more freedom in terms of what we can do, not being limited by the constraints of on-camera signal processing (in terms of speed, algorithm complexity and so on). Offline we may use more sophisticated color constancy methods, which do not require all the simplifying assumptions enumerated before and which can therefore deal more accurately with realistic situations where surfaces are not perfectly diffuse but have specular reflections, where there is more than one illuminant or it is not uniform, etc. We'll discuss some techniques of this kind in Chapter 7. In fact, in [268] it is shown how even small departures from perfectly uniform illumination generate considerable deviations in appearance from reflectance, which goes to say that the diagonal models of white balance, based on discounting the illumination and equating appearance with reflectance, do not predict appearances in real life scenes. But offline white balance requires that the image date is stored as *raw*, i.e. as it comes from the sensor, and many cameras do not have this option, storing the image frames already in color corrected (and demosaicked, and compressed) form.

If offline white balance is not an option, then the next best possibility is that of manual illumination estimation: the operator points the camera towards a reference white, such as a simple sheet of paper, and the triplet of values recorded for this object (at, say, the center of the image) is used by the camera as $(I(\lambda_R), I(\lambda_G), I(\lambda_B))$ to perform the color correction (because in a white object we have that $(S(\lambda_R) = S(\lambda_G) = S(\lambda_B))$, so the observed (R, G, B) should be equal to $(I(\lambda_R), I(\lambda_G), I(\lambda_B)))$. This method ensures that objects perceived as white at the scene will also appear white in the recorded images, and in general that all achromatic objects will appear gray. But cinematographers often find this effect too realistic, and use instead the manual white balance for artistic expression by fooling the camera, giving it as reference white an object with a certain color; in this way, by deliberately performing a wrong color correction, a certain artistic effect can be achieved [143].

Finally there is the option of automatic white balance (AWB), where the illumination estimation is automatically performed on-camera and the color correction (done as in Equation 3.7 or Equation 3.9, i.e. "discounting the illuminant") is carried out on the raw data domain or just after color interpolation [334]. The most common approaches for illuminant estimation in AWB are:

- *Gray World.* This approach was formalized by Buchsbaum in 1980 [112], although the same technique was proposed by Judd forty years before [213, 216]. The main assumption here is that the colors present in the scene are sufficiently varied, in which case the average of reflectances is gray; in other words, that reflectances are uniformly distributed over the interval $[0, 1]$ and hence their average is 0.5 for each waveband. Assuming also that the illuminant is uniform and averaging Equation 3.4 over the whole image, we get:

$$\frac{1}{\mathcal{A}}\sum_{x,y} R(x,y) = I(\lambda_R)\frac{1}{\mathcal{A}}\sum_{x,y} S(\lambda_R,x,y) = \frac{I(\lambda_R)}{2} = R_{average}$$
$$\frac{1}{\mathcal{A}}\sum_{x,y} G(x,y) = I(\lambda_G)\frac{1}{\mathcal{A}}\sum_{x,y} S(\lambda_G,x,y) = \frac{I(\lambda_G)}{2} = G_{average}$$
$$\frac{1}{\mathcal{A}}\sum_{x,y} B(x,y) = I(\lambda_B)\frac{1}{\mathcal{A}}\sum_{x,y} S(\lambda_B,x,y) = \frac{I(\lambda_B)}{2} = B_{average}, \quad (3.10)$$

  where $\mathcal{A}$ is the area of the image. In practice, one channel (usually the green one) is taken as reference and the other two are scaled to perform the color correction, since AWB is only concerned about the ratio of the color signals [239]:

$$
\begin{aligned}
R'(x, y) &= \alpha R(x, y) \\
G'(x, y) &= G(x, y) \\
B'(x, y) &= \beta B(x, y),
\end{aligned} \tag{3.11}
$$

where:

$$
\alpha = \frac{G_{average}}{R_{average}}, \quad \beta = \frac{G_{average}}{B_{average}}. \tag{3.12}
$$

The end result is that, after this correction, the average of the new image is gray: $R'_{average} = G'_{average} = B'_{average}$. Clearly, this method is not effective if the main hypothesis is violated, i.e. if the colors of the image are not sufficiently varied (e.g. when a large, monochromatic object takes up most of the image). To try to overcome this limitation, some approaches like [280] compute the color distribution of achromatic charts under a set of typical illuminants, then decompose the image into blocks and for each block analyze its color differences so as to determine if it's in the same region as that of a common illuminant; finally, all blocks are considered, along with the absolute scene light level, to determine the likely scene illuminant [303].

- *White patch.* This approach is based on the fact that the brightest object in a scene is perceived as white. The observation of this phenomenon is often attributed, incorrectly, to the Retinex theory of Land [243], but it has a long history that dates back at least to the works of Helmholtz [217, 214]. We start by computing the maximum value for each color channel:

$$
\begin{aligned}
R_{max} &= max_{x,y} R(x, y) \\
G_{max} &= max_{x,y} G(x, y) \\
B_{max} &= max_{x,y} B(x, y),
\end{aligned} \tag{3.13}
$$

although it is usually preferred to lowpass the image first or to treat clusters of pixels, so as to avoid the problems caused by a few bright pixels that are outliers [239]. Next we compute the scaling factors:

$$
\alpha = \frac{G_{max}}{R_{max}}, \quad \beta = \frac{G_{max}}{B_{max}}. \tag{3.14}
$$

And finally, the correction is:

$$\begin{aligned} R'(x,y) &= \alpha R(x,y) \\ G'(x,y) &= G(x,y) \\ B'(x,y) &= \beta B(x,y), \end{aligned} \tag{3.15}$$

After the White Patch correction, the brightest point in the scene becomes achromatic: $R'_{max} = G'_{max} = B'_{max}$.

Gray World and White Patch are very commonly used for AWB, sometimes combined, as reported in [150].

Once more we must point out that, as was the case with automatic exposure and automatic focus control, AWB is absolutely avoided by cinema professionals, both for technical and artistic reasons. Technically, AWB produces visible fluctuations in color in dynamic scenes, which is definitely something we don't want. Artistically, the choice of color palette of a shot is a fundamental aspect of the director's vision and therefore it is not something that one would like to automatize.

3.6 Color transformation

3.6.1 The colorimetric matrix

At first glance, it would seem that for the camera to accurately capture colors matching our perception, the color triplets obtained by the camera sensor(s) should correspond to the cone responses of the human visual system. But this is never the case, as Figure 3.12 shows, and since the spectral sensitivities of sensor and cones are different, the responses must also be different. There are several reasons for this difference, like the fact that it is difficult to tune the spectral response of the pigments or dyes of the CFAs, and that having spectral sensitivities with a large amount of overlap (as in the responses of medium and long wavelength cones) would not be practical for the sensor from a signal-to-noise point of view [204]. But while emulating cone responses is not practical for image capture, it is essential in the subsequent processing of the image signal: the stimulus the scene would have produced in the human visual system must be estimated as accurately as possible [204]. This is why we must be able to transform the (R, G, B) values of the sensor into (X, Y, Z) tristimulus values, i.e. go from RGB into CIE XYZ, which we recall is a perceptually-based color space that uses the color-matching functions $\bar{x}, \bar{y}, \bar{z}$ of a standard observer:

$$
\begin{aligned}
X &= \int_{380}^{780} \bar{x}(\lambda)L(\lambda)d\lambda \\
Y &= \int_{380}^{780} \bar{y}(\lambda)L(\lambda)d\lambda \\
Z &= \int_{380}^{780} \bar{z}(\lambda)L(\lambda)d\lambda,
\end{aligned} \tag{3.16}
$$

where L is the irradiance and the color-matching functions are a linear transformation of the cone sensitivities [205].

We can transform (R, G, B) into (X, Y, Z) by imposing the Luther-Ives condition [341]: that the sensor response curves are a linear combination of the color matching functions. We recall from Equation 3.3 that:

$$
\begin{aligned}
R &= \int_{380}^{780} r(\lambda)L(\lambda)d\lambda \\
G &= \int_{380}^{780} g(\lambda)L(\lambda)d\lambda \\
B &= \int_{380}^{780} b(\lambda)L(\lambda)d\lambda,
\end{aligned} \tag{3.17}
$$

where r, g, b are the camera spectral response functions for the red, green and blue channels, so the Luther-Ives condition can be stated as:

$$
\begin{bmatrix} r(\lambda) \\ g(\lambda) \\ b(\lambda) \end{bmatrix} = \begin{bmatrix} b_{11} & b_{12} & b_{13} \\ b_{21} & b_{22} & b_{23} \\ b_{31} & b_{32} & b_{33} \end{bmatrix} \begin{bmatrix} \bar{x}(\lambda) \\ \bar{y}(\lambda) \\ \bar{z}(\lambda) \end{bmatrix}, \tag{3.18}
$$

or, alternatively as:

$$
\begin{bmatrix} \bar{x}(\lambda) \\ \bar{y}(\lambda) \\ \bar{z}(\lambda) \end{bmatrix} = \begin{bmatrix} a_{11} & a_{12} & a_{13} \\ a_{21} & a_{22} & a_{23} \\ a_{31} & a_{32} & a_{33} \end{bmatrix} \begin{bmatrix} r(\lambda) \\ g(\lambda) \\ b(\lambda) \end{bmatrix}. \tag{3.19}
$$

As we just commented above, manufacturing processes and the properties of the materials used make it difficult to adjust at will the sensor response curves, and the Luther-Ives condition is usually not met in practice [205], but despite this fact a three-channel camera with three arbitrary sensor response curves is able to estimate the tristimulus values of an object as long as the object's spectral reflections are always composed of three principal components and they don't change steeply with respect to wavelength [205]. This implies that with a linear transformation we can go from the observed (R, G, B) triplet to its corresponding (X, Y, Z) tristimulus value, plugging Equations 3.19 into Equations 3.16:

$$\begin{bmatrix} X \\ Y \\ Z \end{bmatrix} = \begin{bmatrix} a_{11} & a_{12} & a_{13} \\ a_{21} & a_{22} & a_{23} \\ a_{31} & a_{32} & a_{33} \end{bmatrix} \begin{bmatrix} R \\ G \\ B \end{bmatrix} = A \begin{bmatrix} R \\ G \\ B \end{bmatrix}. \tag{3.20}$$

For each different triplet (R, G, B) we could have a different (and optimal) matrix A, but this would definitely be something very unpractical. A single colorimetric matrix A to be applied to all (R, G, B) colors can be computed in the following way [205]:

1. Build a set of n test patches of representative or important colors.
2. Under controlled conditions, with a known illuminant (e.g. D65), measure the tristimulus values of the patches with a tristimulus colorimeter obtaining (X_i, Y_i, Z_i), $1 \leq i \leq n$.
3. Under the same conditions, use the camera to measure the (R, G, B) values of the patches, obtaining (R_i, G_i, B_i), $1 \leq i \leq n$.
4. A colorimetric matrix A gives an estimated tristimulus value $(\hat{X}_i, \hat{Y}_i, \hat{Z}_i)$ from (R_i, G_i, B_i). A is computed so as to minimize the total visual color difference J, which is a weighted sum of the color differences ΔE (computed for instance in a CIE uniform color space) between the target tristimulus (X_i, Y_i, Z_i) and its estimate $(\hat{X}_i, \hat{Y}_i, \hat{Z}_i)$, for each patch i, $1 \leq i \leq n$:

   $J = \sum_{i=1}^{n} w_i \Delta E(X_i, Y_i, Z_i, \hat{X}_i, \hat{Y}_i, \hat{Z}_i)$,

   where w_i are the weights for the different patches. A can be obtained through least squares minimization, for instance.

In [320] it is noted that finding A by minimizing J as defined just above has the problem that the white point is not preserved, i.e. white in RGB is not mapped to white in the CIE XYZ color space; an additional term can be added to J in order to prevent this [166], and more accurate and robust techniques have also been proposed [104].

The colorimetric matrix is a function of the scene illuminant, so ideally a different matrix should be used for each different scene illuminant the camera is working with [303]. The above process finds the best colorimetric matrix (in terms of minimal error) for the calibration patch set under a given illuminant, and many cameras (most consumer models) use only this one matrix [104]. Some cameras come with several pre-set matrices computed under different illuminations. For instance, using the matrix for fluorescent lighting removes a noticeable green cast that would otherwise be present if we used a matrix computed with a standard illuminant like D65 or D50; other pre-sets may correspond for instance to a "film look" (with de-saturated colors), or may give a very vivid color palette. These pre-set matrices can also be adjusted manually so as to achieve a certain image look, since changing the colorimetric matrix affects hue and saturation (the white point is preserved, though, and color matrix adjustment must not be confused with white balance.)

Broadcast cameras were the first to incorporate the possibility of modifying the colorimetric matrix, so that multiple cameras in live broadcasts could be color matched and no color jumps appeared when switching from one camera to another [69].

3.6.2 A note on color stabilization

We expect two pictures of the same scene, taken under the same illumination, to be consistent in terms of color. But if we have used different cameras to take the pictures, or just a single camera with automatic white balance (AWB) and/or automatic exposure (AE) correction, then the most common situation is that there are objects in the scene for which the color appearance is different in the two shots. Exactly the same happens in video: shots from two different cameras don't match in terms of color, and single-camera video with AE and/or AWB on will show noticeable changes in color whenever the camera motion makes the background luminance change [159].

This is problematic in many contexts. With a single camera, the only way to ensure that all pictures of the same scene are color consistent would be to save images in the RAW format, or to use the same set of manually fixed parameters for all the shots. These are not common choices for amateur users, but professional users face the same challenges: the most popular DSLR cameras for shooting HD video don't have the option of recording in RAW [143]; and while in cinema the exposure and color balance values are always kept constant for the duration of a take (i.e. AE and AWB are never used), the shooting conditions may require us to change these values from shot to shot. With different cameras the problem is aggravated, because using the same parameter values in all cameras is not enough to guarantee the stability of color across shots [270]. In many professional situations several cameras are used at the same time (e.g. large photo shoots, many mid-scale and most large-scale cinema productions), and in some cases the multi-camera set-up is required, not optional (TV broadcasts, 3D cinema).

The end result is that color discrepancies are unavoidable. We can expect most amateur video and stills to exhibit unpleasing color fluctuations, and in the TV and film industries much care and work is devoted to removing these color changes, both in production and post-production. For instance, TV broadcasts employ devices called Camera Control Units (CCU), operated by a technician called a Video Controller or Technical Director (TD): while each camera operator controls most of his/her camera functions such as framing or focus, the TD controls the color balance and shutter speed of a set of cameras so as to ensure color consistency across them, so for instance a 20-camera broadcast may have five CCU operators, each controlling four cameras [376]. In cinema and professional video, both for single and multi-camera shoots, color consistency is a key part of a post-production stage called color grading, and it's performed by seasoned technicians who manually tune parameters so as to match colors between frames [310]. In professional 3D movie production it is crucial that the twin cameras used in a stereo set-up perform in exactly the same way [270]; if there are discrepancies they are typically fixed in post-production by color-matching one view to the other, called the master view, which is taken as reference.

In all digital cinema (not necessarily 3D), the cinematographer's onset looks and settings may not be accurately relayed when that image needs to be corrected by someone else. Because mismatched colors create a huge amount of extra work, in a professional setting cameras must be manually calibrated using a standard color chart, a waveform monitor and a vectorscope [69].

The color chart is a printed reference pattern, as the one shown in Figure 3.13. Charts are also essential when for some reason or another the cameras cannot be color calibrated at the scene: shooting a few frames of the chart at the beginning of each set-up will allow a colorist to properly match across shots in postproduction [69].

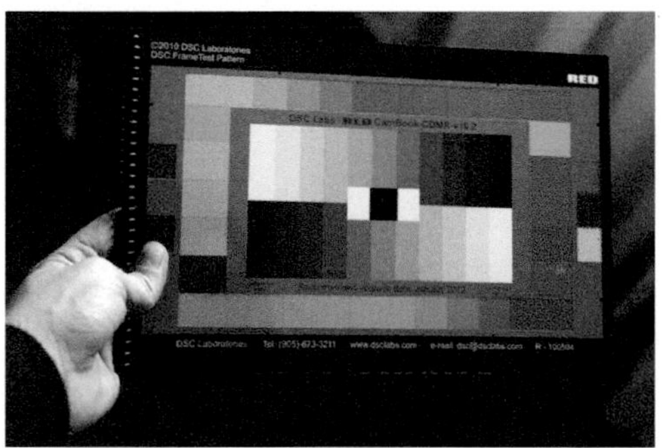

FIGURE 3.13: Color chart. Image courtesy of Ben Cain and Negativespaces.com.

A waveform monitor is an oscilloscope and displays the video signal as a function of voltage with respect to time. With it we can observe the general characteristics of the video signal, especially the black and white levels of the picture (which we can use to adjust the exposure so as to maximize the signal range), and also detect problems such as uneven lighting: if the lighting on the chart is even, the waveform plot should be a horizontal line, otherwise if the plot "dips" this means that the position of the light source must be adjusted; see Figure 3.14.

The vectorscope displays the chrominance of the video signal in a 2D plot, where the center corresponds to white, the angle from the horizontal defines the hue, and the distance from the center indicates saturation. In Figure 3.15 the camera is adjusted so as to maximize saturation for a set of six chart colors.

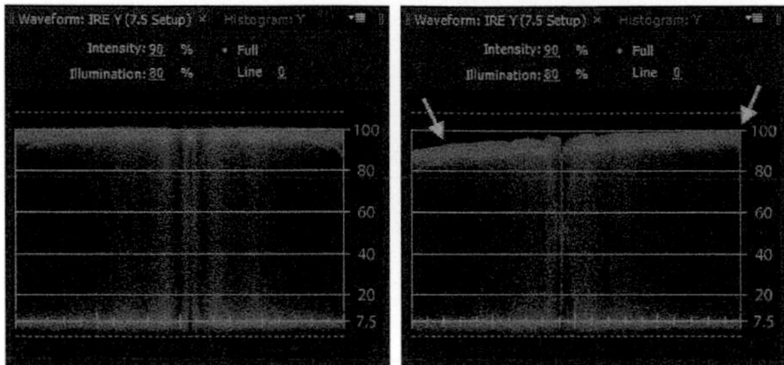

FIGURE 3.14: Waveform plots. Left: even lighting. Right: uneven lighting. Images courtesy of Wayne Norton and DSC Labs.

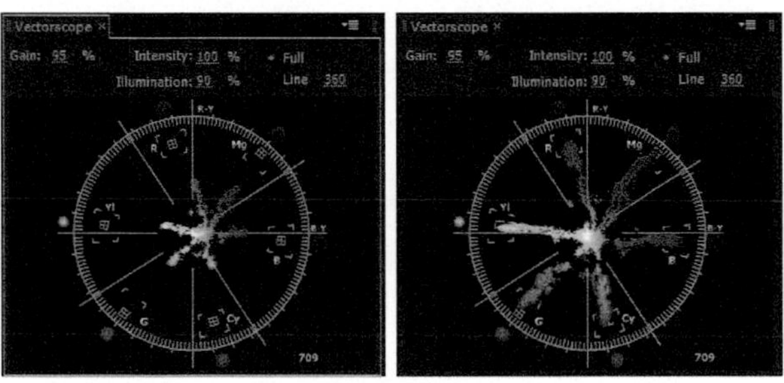

FIGURE 3.15: Vectorscope plots. Left: representation of chart before adjustment (out-of-the-box camera presets). Right: after adjustment. Images courtesy of Wayne Norton and DSC Labs.

3.6.3 Encoding the color values

Now that we have the (X, Y, Z) tristimulus values, which are an estimate of the stimulus the scene would have produced in the human visual system, we must convert them to a standard RGB color space so that the image can be displayed by a device like a TV monitor, a computer screen or a digital projector. Within a given color imaging system, the encoding provides a link between the input obtained by the camera and the output of the display [184]. We can't just use the original (R, G, B) values and omit the conversion to (X, Y, Z), because the output device has its own RGB color space which is not related to and does not match the color space of the camera. We'll briefly review these concepts in the following paragraphs.

Recall that in CIE xy coordinates, monochromatic colors (corresponding to light of a single wavelength) lie on the boundary of the horseshoe-shaped region which corresponds to the visible spectrum. Colors that are mixtures, i.e. corresponding to light with a power spectrum that covers different wavelengths, lie inside this region. As a color becomes more mixtured and its spec-

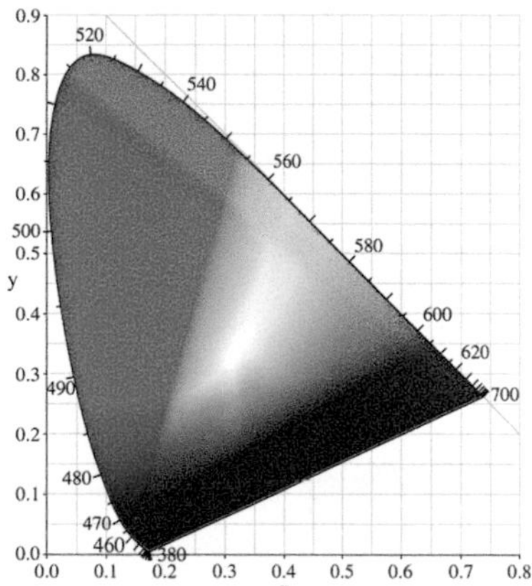

FIGURE 3.16: The color gamut of the visible spectrum in CIE xy coordinates. Image from [13].

trum spreads, its xy location in the diagram moves inward, with achromatic colors in the center; see Figure 3.16.

In a trichromatic device like a TV display or a digital projector, colors are created as a linear combination of the three primaries of the device, the three particular primary colors it uses. Therefore, if we want to represent in CIE xy coordinates the color gamut of a device (the set of colors it can reproduce), it will be the triangle with vertices in the points $(x_R, y_R), (x_G, y_G), (x_B, y_B)$ which are the xy coordinates of the R, G and B primaries. All colors reproducible by the device will lie inside this triangle, because they are a linear combination of $(x_R, y_R), (x_G, y_G)$ and (x_B, y_B). These primaries are (except for very special hardware) not monochromatic, because the physical characteristics of the materials with which these devices are manufactured make them have spectral sensitivities that are rather spread out. Therefore the primaries lie not on the boundary but inside the visible spectrum region, and the implication is clear: there are many colors that we could see but that the device isn't capable of reproducing. For instance, Figure 3.17 shows the color gamut of a CRT television set, where the primaries are given by the spectral characteristics of the light emitted by the red, green and blue phosphors used in CRT's.

In short, the primaries define the color space of the device. Figure 3.18 compares the color gamut of a specific model of digital cinema camera with that of print film, the DCI P3 standard for a digital projector (labeled DCI) and the BT.709 standard for HDTV.

The very wide gamut of the camera shown in Figure 3.18 is due to the dyes used in the CFA, while the same camera maker uses for another of its models a different set of dyes so that its color gamut is BT.709 [2], much smaller. An

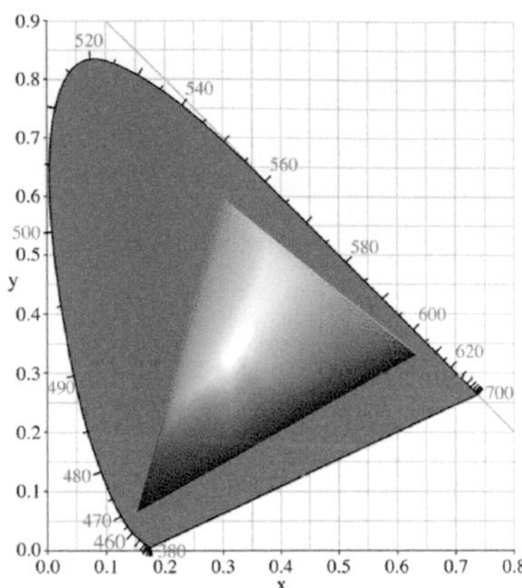

FIGURE 3.17: The color gamut of a CRT television set. Image from [12].

obvious question then would be: why doesn't the maker use the wide-gamut CFA for all its camera models? The answer is that depending on the intended use and market of the camera, too much color space from the sensor may not be a good thing. If the camera will be used in broadcasting to be screened on regular monitors (which adhere to the BT.709 standard), then a wide-gamut camera is wasting processing and information space on colors that will never be reproduced; on the other hand, for digital cinema work the goal is not to have the camera be the limiting factor so that current and future display mediums will best be served by the cameras capabilities, in which case the widest gamut is desirable [2].

Furthermore, for a given intended output (e.g. HDTV) there are also differences in gamut for different technologies. For instance, Figure 3.19 shows the gamuts of TV monitors based on CRT, LCD and plasma technology.

This brings us back to the issue of how to encode the color information. Given that the primaries define the gamut and that they vary greatly among devices, it is essential that the camera uses a standard set of primaries for color encoding, so that each display device can adapt later on to these standard primaries. This is the usual arrangement in color encoding for all imaging systems, splitting the color processing into two parts, one for the input and another for the output so that each has its own associated transform, *to* and *from* a previously agreed upon color encoding specification [184].

The main possibilities then for primary sets are three, corresponding to the following gamuts:

- BT.709 (ITU-R Recommendation BT.709, also known as Rec 709). Agreed upon in 1990, it defines the HDTV standard. For technical reasons (compatibility and noise reduction) it uses primaries very close to the phosphor primaries of CRTs, and since newer display technologies

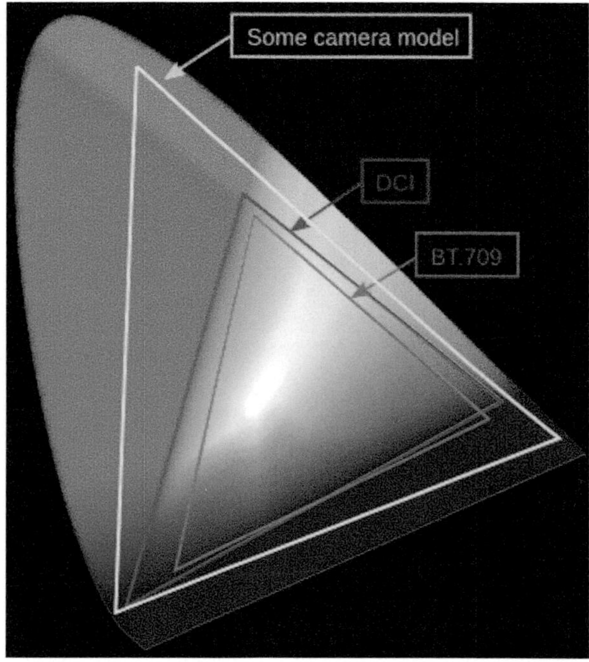

FIGURE 3.18: The color gamut of a digital cinema camera model and of some standards.

are not based on CRTs it is now recognized that color correction at the receiver should compensate for the difference [341]. This is the standard used in broadcast and cable TV, DVD and Blu-ray discs.

- DCI P3 (defined in the Digital Cinema System Specification [34] and in SMPTE 431-2). This is a color gamut for digital cinema projectors, agreed upon in 2006. It is based on the gamut of Xenon lamps which are commonly used in digital projectors, approximating (and in some cases surpassing) the color gamut of film, and therefore it is wider than BT.709. Professional movies today are typically mastered in DCI P3 [314]. Figure 3.20 compares DCI P3 with BT.709.
- DCI $X'Y'Z'$ (defined in the Digital Cinema System Specification [34]). Digital Cinema Initiative (DCI) is a consortium established by major motion picture studios, formed to develop a standard architecture for digital cinema systems. After considering several options for the primaries, such as a wider-gamut RGB space that encompassed all film colors or a parametric RGB encoding with the primaries of the projector as metadata, it was pointed out that rather than argue over which set of wide gamut primaries to use, DCI could just adopt the widest gamut set:

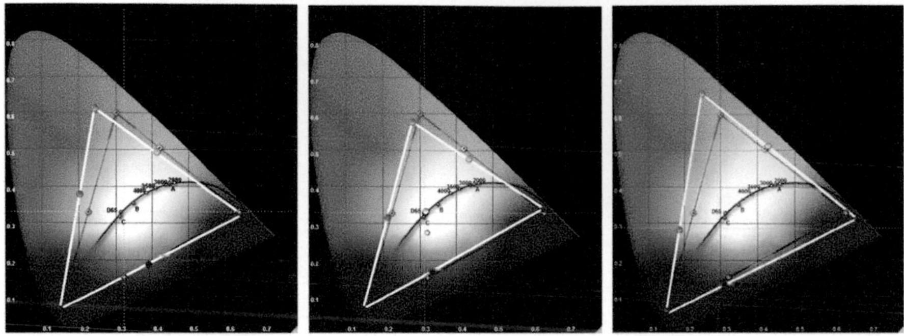

FIGURE 3.19: The color gamut (as white triangles) of different TV display technologies, with reference to the BT.709 color space (black triangle). From left to right: CRT, LCD, plasma. Images by P.H. Putman [317].

> the CIE XYZ primaries [226]. These primaries are the basis of a widely used international standard and met all DCI requirements, specifically providing no limits on future improvements since they enclose all visible colors. Since these primaries fall outside the visible spectrum they are not real but virtual primaries. Several movies have been encoded and released in this format [226]. Figure 3.21 compares DCI $X'Y'Z'$ with DCI P3.

Given the chromaticity coordinates $(x_R, y_R), (x_G, y_G), (x_B, y_B)$ of the primaries of an RGB system, and also the (X_w, Y_w, Z_w) value of its white point, the camera converts the XYZ tristimulus values computed in the previous section to output RGB values (compliant with BT.709 or DCI P3, for example) by multiplying by a 3x3 matrix [32]:

$$\begin{bmatrix} R \\ G \\ B \end{bmatrix} = M^{-1} \begin{bmatrix} X \\ Y \\ Z \end{bmatrix}, \tag{3.21}$$

where:

$$M = \begin{bmatrix} S_r X_r & S_g X_g & S_b X_b \\ S_r Y_r & S_g Y_g & S_b Y_b \\ S_r Z_r & S_g Z_g & S_b Z_b \end{bmatrix} \tag{3.22}$$

$$X_r = \frac{x_r}{y_r} \;; X_g = \frac{x_g}{y_g} \;; X_b = \frac{x_b}{y_b}$$
$$Y_r = Y_g = Y_b = 1$$
$$Z_r = \frac{1 - x_r - y_r}{y_r} \;; Z_g = \frac{1 - x_g - y_g}{y_g} \;; Z_b = \frac{1 - x_b - y_b}{y_b} \tag{3.23}$$

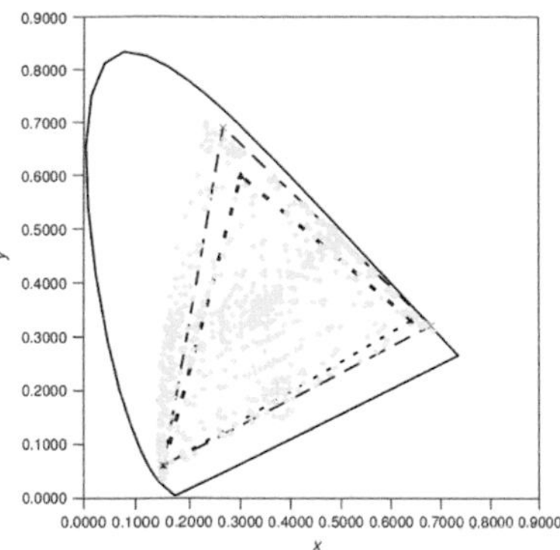

FIGURE 3.20: Comparison of color gamut of film (pastel green dots), DCI P3 (dashed red line), BT.709 (dashed blue line). Image from [226].

$$\begin{bmatrix} S_r \\ S_g \\ S_b \end{bmatrix} = \begin{bmatrix} X_r & X_g & X_b \\ Y_r & Y_g & Y_b \\ Z_r & Z_g & Z_b \end{bmatrix} \begin{bmatrix} X_w \\ Y_w \\ Z_w \end{bmatrix}. \tag{3.24}$$

If any of the obtained (R, G, B) values fall outside the $[0, 1]$ interval they are normally *clipped* by the camera; this imposes on the color encoded images the color gamut determined by the primaries $(x_R, y_R), (x_G, y_G), (x_B, y_B)$, and it's a lossy operation which can't be undone. Therefore it is preferred that the camera performs color encoding in the widest gamut it is capable of, minimizing information loss. Then after post-production this gamut can be adapted to different output gamuts in a color correction stage performed by a skilled colorist, who is sensitive to the creative intent of the movie and who is under the supervision of the director and/or cinematographer [314]; this last stage of color correction, just prior to movie distribution, usually involves 3D LUTs [86]. In the academic literature the problem of adapting color gamuts is called *gamut mapping* [282] and we discuss it in Chapter 10.

3.7 Gamma correction and quantization

3.7.1 The need for gamma correction

At the onset of broadcast television it was observed that CRTs produce luminance as a non-linear, *power* function of the device's voltage input: $L = \alpha V^\gamma$, where L is the luminance, V the voltage and γ the exponent of this power function, which has a value of around 2.5 (α is just a proportionality

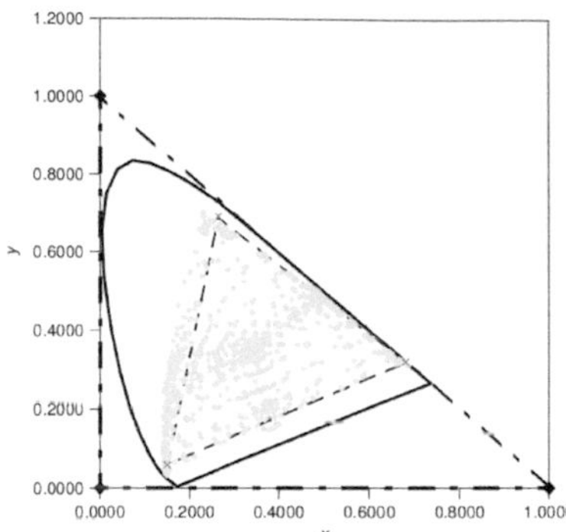

FIGURE 3.21: Comparison of color gamut of film (pastel green dots), DCI P3 (dashed red line), DCI $X'Y'Z'$ (dashed black line). Image from [226].

coefficient). But the luminance signal captured by the camera is linearly proportional to light intensity. Therefore, for correct luminance reproduction on a TV set, the camera's luminance signal must be non-linearly scaled with a power function with the inverse exponent of the CRT's power function, thus both non-linearities cancel each other out and the luminance of the CRT can be a faithful (scaled) representation of the luminance reaching the camera: $V = \beta L'^{\frac{1}{\gamma}}$, where L' is the camera's luminance (β is just a proportionality coefficient). This process is called *gamma correction*. See Figure 3.22.

At that time it was also very well known that humans have a perceptual response to luminance that is also non-linear: perceived lightness is roughly the 0.42 power of physical luminance [313]; see Figure 3.23. This means that differences in the dark parts of an image are more noticeable than differences *of the same amount* on bright parts of the same image. Given that over-the-air analog transmission introduced noise in the TV signal, a simple linear transmission of the luminance values captured by the camera would make this noise much more apparent in the darkest regions of the image. Therefore, non-linearly scaling the camera's luminance with a power function of exponent 0.42 makes this noise less perceptible.

The amazing coincidence, as Charles Poynton puts it in [313], is that the CRT voltage-to-luminance function is very nearly the inverse of the perceptual luminance-to-lightness relationship, i.e. the lightness perception curve of Figure 3.23 is very similar to the power function of exponent $\frac{1}{\gamma}$ that compensates for the CRTs nonlinearity (see Figure 3.22), a fact already recognized in 1939 [314]. In the early days of TV, then, gamma correction was essential for two tasks: compensating for the CRT nonlinear response, and reducing noise through perceptual encoding, which implies that even if the CRT response had been linear the gamma correction process would have had to be performed in the same way.

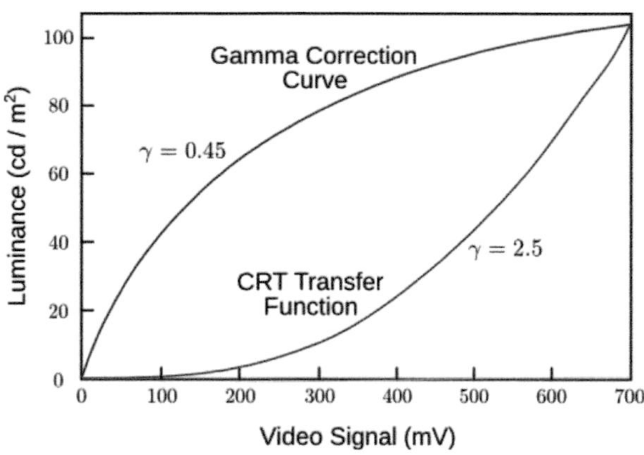

FIGURE 3.22: Gamma correction on a CRT.

But why is gamma correction still used today, when CRTs have become obsolete and air transmission noise is no longer a major concern with the advent of digital TV?

The main reason is that since digital signals have a limited number of bits to code each pixel value, if we quantize a gamma corrected signal the quantization intervals are wider at higher luminance values, where changes are less perceptible. In other words, gamma correction allows us to use more bits at the darkest regions, where we are more sensitive to differences, and less bits at the brightest regions, where we are less sensitive to differences; this is what we mean by perceptual coding. For a fixed number of bits, then, perceptual coding permits us to maximize image quality (or, conversely, without perceptual coding we would have to use more bits to represent images with the same perceptual quality).

There is another, very important reason, which is the issue of color appearance. As explained in [314], we can't just aim for the reproduced image to have values proportional to those in the original scene, because the appearance of the images is modified by the environment in which they are seen by the viewer; this environment is usually different from that of the original scene, typical displays have lower luminance and contrast than typical scenes, and their surrounds are darker. The differences in environment produce three main effects that must be compensated [314]:

- The Hunt effect: colorfulness decreases as illumination decreases. If an image is taken in daylight and linearly displayed in a dim environment, the image will look as if it was captured at twilight. See Figure 3.24.

- The Stevens effect: contrast decreases as illumination decreases, i.e. dark colors look lighter and light colors look darker. See Figure 3.24.

- The simultaneous contrast effect: the center squares in Figure 3.25 have

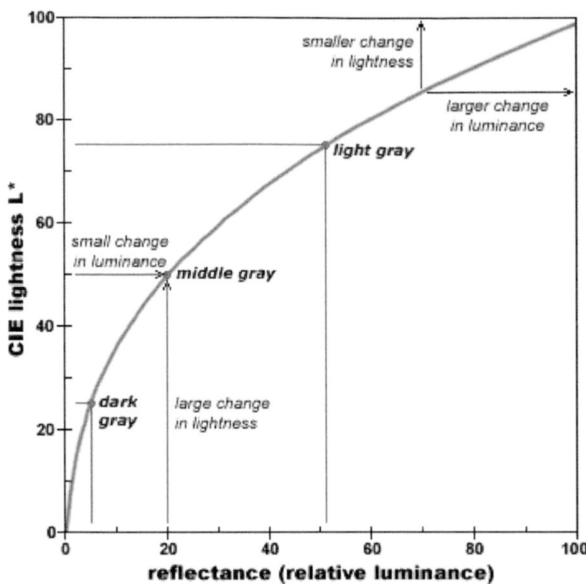

FIGURE 3.23: Lightness perception as a function of luminance is approximately a power function of exponent 0.42. Image from [59].

the same shade of gray, but the one surrounded by dark seems lighter than the other center square.

Experience shows that all three appearance effects can be ameliorated at the same time by imposing a modest end-to-end power function: rather than encoding with a power $\frac{1}{\gamma}$ and decoding with γ, different gamma values are used for encoding (γ_E) and decoding (γ_D) so that the net effect is a power function with an exponent slightly greater than one, $\frac{\gamma_D}{\gamma_E} \simeq 1.2$ [314].

For the aforementioned reasons, cameras normally apply gamma correction, which is performed after the color correction stage (some high-end digital cameras allow recording of a linear output, without gamma correction). The actual value of gamma that is used depends on the intended viewing conditions. If using the BT.709 color encoding, intended for HDTV viewing on low contrast displays in dim conditions, the encoding gamma is equivalent to $0.45 = \frac{1}{2.2}$. If using DCI P3 or DCI $X'Y'Z'$ color encoding, intended for viewing in a cinema with dark surround and high contrast, the encoding gamma is $\frac{1}{2.6}$. In the BT.709 case the end-to-end coefficient of the power function is 1.2, while in the digital cinema case it is 1.5 [314].

It must be noted though that cameras also use power-laws to enhance contrast and achieve pleasant-looking images, aside from the gamma-correction inherent to the standard [228]. The end result is that the actual gamma value of any recorded shot may not be the gamma value of the standard the shot has

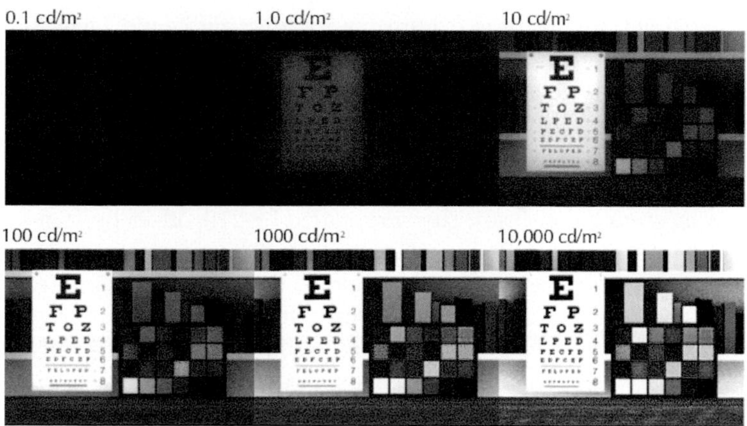

FIGURE 3.24: Hunt and Stevens effects. Image from [155].

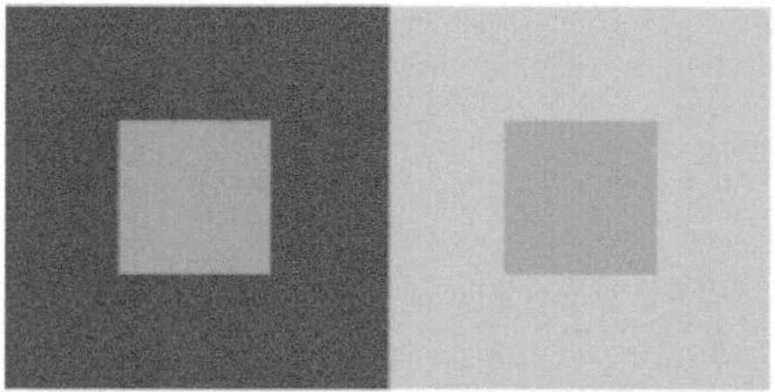

FIGURE 3.25: The simultaneous contrast effect.

been recorded in; if this value is needed (e.g. to perform linearization previous to color correction) it needs to be estimated.

3.7.2 Transfer function and quantization

In order to implement gamma correction a transfer function is used. For instance, in BT.709, each R, G and B component is processed with the following transfer function, obtaining a gamma-corrected value R', G' or B':

$$V' = \begin{cases} 4.5L, & 0 \leq L \leq 0.018 \\ 1.099L^{0.45} - 0.099, & 0.018 \leq L \leq 1 \end{cases} \tag{3.25}$$

where L denotes R, G or B and V' the corresponding gamma-corrected value. Notice how this is not a pure power function, it has a linear segment near black, with a limited slope. The reason for this is practical: the slope of a pure power function whose exponent is less than one is infinite at zero, therefore a transfer function that was a pure power function would have infinite gain near black [313]. Poynton [313] points out that because of the linear segment introduced for low values, the overall transfer function is very similar to a square root ($\gamma_E \simeq 0.5$), hence it is not accurate to describe BT.709 as having $\gamma_E = 0.45$.

The values in Equation 3.25 are in the range $[0, 1]$. For quantization in 8 bits (the most common case), they are scaled by 219, offset by 16 and rounded to nearest integer; values below 16 and above 235 are reserved [313] and provide range for filter over and under shoots. Digital video standards were developed as an addition to analog broadcast standards and were required to be backward-compatible with existing analog video equipment, hence providing ample headroom for analog variations was a prudent practice to avoid clipping artifacts [226]. For quantization in 10 bits, scaling and offset values are multiplied by 4. Digital still cameras encode pictures in sRGB, which utilizes the full range $0 - 255$.

For digital cinema, the DCI specification [34] states that the transfer function for the XYZ tristimulus values must be calculated with a normalizing constant of 52.37 (corresponding to an absolute luminance of $52.37cd/m^2$) and a gamma value of 2.6:

$$CV_{X'} = round\left(4095 \times \left(\frac{X}{52.37}\right)^{\frac{1}{2.6}}\right) \tag{3.26}$$

$$CV_{Y'} = round\left(4095 \times \left(\frac{Y}{52.37}\right)^{\frac{1}{2.6}}\right) \tag{3.27}$$

$$CV_{Z'} = round\left(4095 \times \left(\frac{Z}{52.37}\right)^{\frac{1}{2.6}}\right) \tag{3.28}$$

The DCI specification establishes quantization in 12 bits per component, following the results from the experiments in [131]. These experiments were performed in cinema viewing conditions and demonstrated that 12 is the minimum number of bits necessary for observers not to perceive quantization artifacts. Unlike with BT.709, the full range is used and there are no illegal code values in digital cinema mastering and distribution, since they are both purely digital processes that therefore don't require us to provide headroom for analog variations [226]. Nevertheless, when digital systems use the SMPTE 372 dual-link high definition serial interface (HD-SDI) they are subject to its reserved code values, which are $0 - 15$ and $4080 - 4095$: values below 15 or above 4080 are clipped, although in practice this is visually unnoticeable [226].

3.7.3 Color correction pipeline

We can summarize all the above elements of the color processing chain as follows [103]:

$$\begin{bmatrix} R \\ G \\ B \end{bmatrix}_{out} = \left(\alpha \begin{bmatrix} c_{11} & c_{12} & c_{13} \\ c_{21} & c_{22} & c_{23} \\ c_{31} & c_{32} & c_{33} \end{bmatrix} \begin{bmatrix} r_{AWB} & 0 & 0 \\ 0 & g_{AWB} & 0 \\ 0 & 0 & b_{AWB} \end{bmatrix} \begin{bmatrix} R \\ G \\ B \end{bmatrix}_{in} \right)^{\gamma} \quad (3.29)$$

where RGB_{in} is the camera raw triplet, to which a diagonal white balance matrix is applied, followed by the matrix $[c_{ij}]$ which cascades the colorimetric matrix of Equation 3.20 with the color encoding matrix of Equation 3.21, a gain factor α and finally a power function of exponent γ is applied. RGB_{out} is the output of this color correction pipeline but not the actual triplet value recorded by the camera, because as we shall see there still remain some image processing operations in the full camera pipeline, e.g. edge enhancement and video compression, which will alter the final values.

In [303] it is pointed out that the RGB_{in} camera raw triplet is usually not the original sensor signal but a corrected version of it, where the original nonlinear camera exposures have been linearized through a LUT; thus, we can assume that all RGB_{in} triplets are actually proportional to the exposures. The final gamma correction power function is also implemented with a LUT, and since the color transformation matrices can be cascaded into a single 3x3 matrix, the whole color processing pipeline can be expressed as a LUT-matrix-LUT sequence [303].

3.8 Edge enhancement

As we saw in a previous chapter, the optical properties of the lens system determine its MTF, which expresses contrast as a function of spatial frequency. The lens system produces image blurring, making the MTF decrease faster, but so do the optical anti-aliasing filter and the sensor aperture. The result is that the sharpness of the image is reduced, because perceived sharpness can be approximately quantified as the area under the MTF curve squared, the "equivalent line number" proposed in 1948 by Otto Schade [31]. In order to compensate for this loss of contrast, many cameras incorporate an edge enhancement process, also called sharpening or "unsharp masking." Because of perceptual considerations already discussed, the amount of sharpening to be performed depends on the viewing conditions (mainly screen size and viewing distance, but also the luminance of the surround). Because these are variable, digital camera edge enhancement is usually optimized for a selected and conservative viewing condition [68]. As with other automated processes mentioned before in this chapter (e.g. AWB), cinematographers prefer that the camera doesn't perform edge enhancement and instead prefer to boost contrast in post-production where there is much more control over the process and a wide range of possible algorithms to choose from. In Chapter 7 we'll see color correction methods that also enhance the contrast without introducing artifacts such as halos or spurious colors.

Basic edge enhancement is linear: an edge map E is computed from the image I, scaled and added back to I to obtain the sharpened image I' [68]:

$$I' = I + kE. \tag{3.30}$$

A common way to compute the edge map is through an "unsharp mask": the image I is blurred by convolution with a Gaussian g, then subtracted from the original I to obtain E [68]:

$$E = I - g * I. \tag{3.31}$$

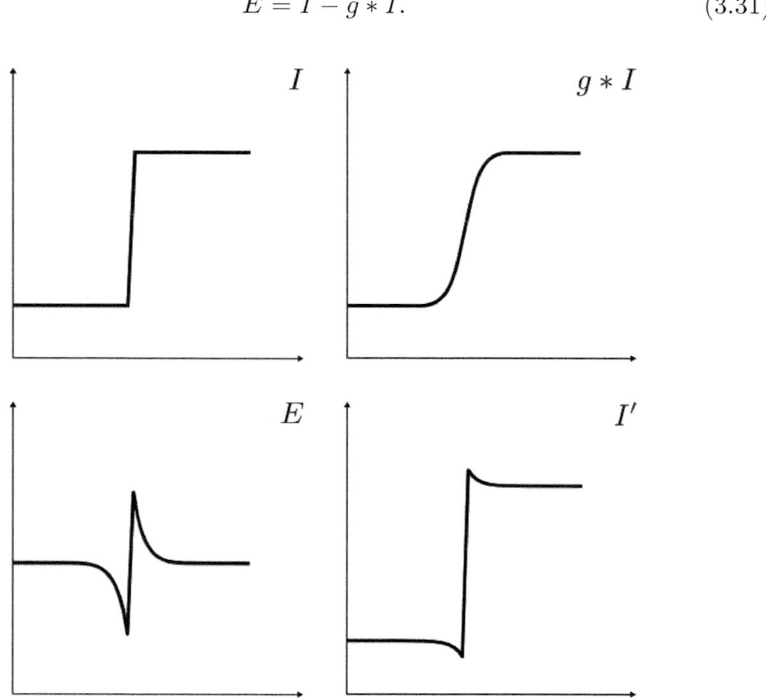

FIGURE 3.26: Linear edge enhancement (unsharp masking). Top left: edge. Top right: aftter Gaussian blur. Bottom left: edge map. Bottom right: sharpened result.

The term "unsharp masking" comes from film photography and originally referred to the use of an unsharp (blurred) positive film mask made from the negative: when this mask was contact printed in register with the negative, the contact print would accentuate the edges present in the image, resulting in an image with a crisper and sharper look [5]. We can see that this process, translated into digital terms, is exactly the same as what Equations 3.30 and 3.31 are doing. Figure 3.26 shows, on the top left, an edge plot from the original image (I); on the top right, the image after blurring ($g * I$); on the bottom left, the edge map (E); and on the bottom right, the sharpened result (I'). This figure shows how the resulting edge has been enhanced, but it is also clear that if the scale parameter k in Equation 3.30 is too large the enhancement's over and undershoots may be perceived as halos, which is definitely something to be avoided. Better results can be obtained with nonlinear edge enhancement. It consists of applying to the edge map E a nonlinear function such as the one depicted in Figure 3.27, for instance through a simple LUT operation; this soft-thresholding operation reduces noise by eliminating edge enhancement for small values, and reduces halos by limiting large edge values [68].

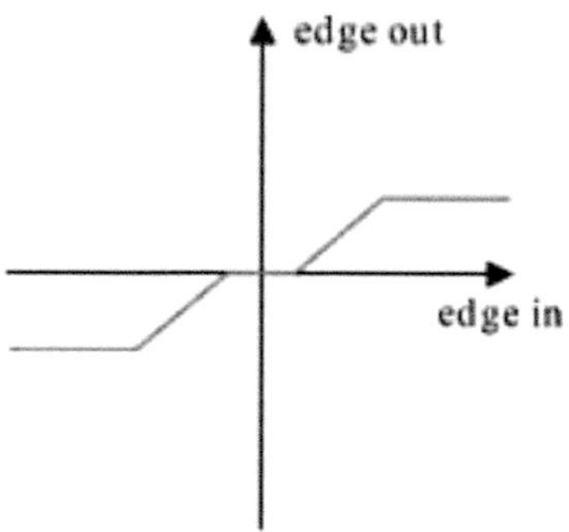

FIGURE 3.27: Nonlinear edge enhancement function. Figure from [68].

Though it would seem preferable to perform edge enhancement in a color space that is linear with exposure, since many optical degradations such as optical blur occur in linear space, experience shows that it is actually better to do the sharpening on a color space that is uniform with visual perception, which is why in-camera edge enhancement is typically performed after gamma correction [303]. Also, edge enhancement is often applied to the luma channel (the Y channel of the YCbCr space) instead of the RGB color channels, so as to avoid amplification of colored edges created by chromatic aberration or other artifacts earlier in the image processing pipeline [68].

3.9 Output formats

Recording images of 2880x1620 pixels at 30 frames per second and with 10 bits of color information per channel requires a data rate of approximately 4 gigabits per second. This figure is beyond the capabilities of todays' cameras, which must therefore resort to compression before recording in-camera; a few models, like the ARRI Alexa, allow for uncompressed recording but this must be performed off-camera and requires special recording equipment.

3.9.1 Compression

The compression may be lossless, in which images can be recovered without any information loss, or lossy, with information loss increasing with the compression rate.

Lossless compression is based on Huffman coding and run-length encoding and is normally used for raw data, as in the CinemaDNG format, based on the TIFF/EP format with which raw images are recorded in several models of still cameras.

For lossy compression there are many alternatives. For starters, the color coding can be in RGB or YCbCr. In YCbCr there are different possibilities of color subsampling, noted thus:

- 4:4:4, no subsampling. This is the preferred choice for digital cinema.
- 4:2:2, reduction of the chroma components by a factor of 2 in the horizontal direction. Common in HDTV cameras for broadcast.
- 4:2:0, reduction of the chroma components by a factor of 2 in both the horizontal and vertical directions. Common in reflex still cameras capable of shooting HD video [143].

Next, there is the possibility of using intra-frame coding or predictive coding.

When each frame is compressed independently from the others, treating them as standalone images, then all frames are said to be of type I for "intra-coded," coded by intra-frame techniques. This is the approach used in several formats such as ProRes, DNxHD, and REDCODE.

Movies have a very high time redundancy, so if we want to increase the compression rate we may want to code frame differences instead of the frames themselves: for slowly varying sequences, the differences among motion-compensated images will be small and hence can be coded using less bits. This sort of coding, where only "new" information is processed, is called predictive coding. Frames encoding (motion-compensated) differences with another frame are called type P for "predictive-coded," or B for "bi-directionally predictive-coded." The image sequence is partitioned into groups of pictures, GOPs, which may last from 1 to 15 pictures or more. Each GOP is independent from the others: all P and B frames in a GOP code motion-compensated differences with other I and P frames in the same GOP. A GOP always starts with an I frame, so if a codec only uses one-frame GOPs it is in fact coding all frames with intra-coding and all frames are independent. Videos in which GOPs have more than one picture are called Long-GOP. Very popular formats like MPEG-1, MPEG-2 and MPEG-4/AVC H.264 use predictive coding, i.e.

their frames may be of type I, P or B. If a video in one of these formats uses Long-GOP it can't be edited directly due to the presence of prediction-coded frames P or B; it must be converted to a format suitable for editing in a process called "rushes upgrade," going for instance from H.264 to ProRes 422 [143].

Finally, there's the actual lossy compression technique used, which can be based either on the DCT (discrete cosine transform) or on the wavelet transform. The DCT is the basis for MPEG-1, MPEG-2, MPEG-4/AVC H.264, DNxHD and ProRes (and for the JPEG format for still images). The wavelet transform is the basis for REDCODE and DCI $X'Y'Z'$ [34], both of which use only I frames coded with variants of JPEG2000 (a wavelet-based compression standard for still images). Thresholding the transform coefficients is a typical denoising technique known as *coring* [320]: coefficients below the threshold are assumed to correspond to noise and set to zero. This is a basic denoising technique, as we mention in Chapter 5.

We will see compression in some more detail in Chapter 4.

3.9.2 Recording in RAW

Some cameras allow for recording in a raw image format, where the data is stored as it comes from the sensor, without the usual in-camera image processing described in the precedent sections: no demosaicking, color balance, color encoding, sharpening, denoising, etc. Without those operations the image is not ready for display, which is what the term *raw* implies. But RAW images present many clear advantages, allowing for the best image quality in digital cinema:

- The original image signal, as captured by the sensor, is recorded, preserving the native 12 bit or 14 bit depth without quantization.
- The original color gamut is also preserved and therefore the output color space can be chosen at will.
- With a RAW image the cinematographer has total freedom in the particular choice of image processing algorithms to be applied in postproduction: for instance, he/she may choose the best demosaicking technique, a custom non-local white balance algorithm, an image-based gamut mapping method, a state of the art non-local denoising algorithm, etc. None of these options are available in current digital cinema cameras, not even high-end professional models.
- Potentially problematic in-camera image processing, such as denoising and sharpening, is avoided.
- Compression artifacts are avoided as well: raw image formats are uncompressed (like ArriRaw), or use lossless compression (like CinemaDNG), or high quality lossy compression (like REDCODE).

– The dynamic range is better preserved than in formats where images are already tone-mapped, gamma-corrected and quantized.

Cinematographer John Brawley expresses very clearly the advantages of shooting in RAW [23]:

"Shooting on film means you could choose your individual film stock with its own personality and then its own developing processes and even what kind of telecine of scanning you wanted to do.
You could choose to alter steps in the chemical baths that meant you could affect contrast (bleach bypass). You could choose if you wanted the Cintel look (flying spot scanner) or the Spirit look (CCD scanner with a soft light source). You could choose how much you wanted to sharpen, how to set up your image and there was always near limitless control of the image.
Shooting RAW is a bit like the electronic equivalent of shooting film because you get to choose a great deal of how the image is processed in post later instead of letting the camera make those decisions for you in a generic way when its being recorded.
RAW means you get a file that has barely any of the usual image processing applied to it that a lot of other cameras normally get. This is really good. And if youre not ready to step up and learn how to deal with it properly, then you wont be getting the most out of the camera.
I like to think of RAW images as the raw ingredients you need to cook a great meal. More processed images like the ones you get out of a 5Dmk2 are kind of like eating out where the cooking is done for you. Sometimes you can get great results, and sometimes youre only going to get fast food. And eventually, you get to missing a home cooked meal."

3.10 Additional image processing

3.10.1 Lens spatial distortion correction

Optical aberrations such as pincushion and barrel distortions can be corrected through in-camera image processing. The distortion amount must be estimated for a given lens, and approximated by a function of the image height from the lens center [334]. See Figure 3.28.

3.10.2 Lens shading correction

Many lens systems produce a vignetting effect on the image, causing the periphery to be darker than its center. The compensation algorithm consists then of amplifying each pixel value depending on the pixel's distance to the image center, and it can be easily implemented with a LUT operation [334];

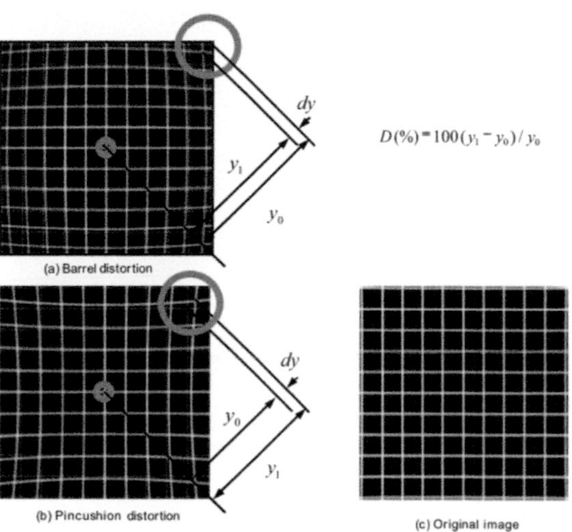

FIGURE 3.28: Lens spatial distortion. Figure from [334].

see Figure 3.29. Some camera models have tens of precomputed correction LUTs for different lens systems and even allow to manually add some more [143].

3.11 The order of the stages of the image processing pipeline

The stages discussed in the preceding sections appear in different order in different pipelines. In [223], Kao et al. propose an image processing pipeline and justify the ordering of its stages based on several considerations:

1. White balance should be performed early in the pipeline, because later operations depend on a correct relationship/ratio among channels.

2. Demosaicking should also preceed other image processing operations where tri-stimulus values are needed.

3. Noise is produced in RGB space, so it's there where it should be addressed, not in YCbCr, where many denoising algorithms work (they do so because the human visual system is more sensitive to noise in the luminance channel than in the chroma ones). Also, noise reduction should come before color correction, because this latter process may enlarge random noise. And it would be best if the denoising procedure would

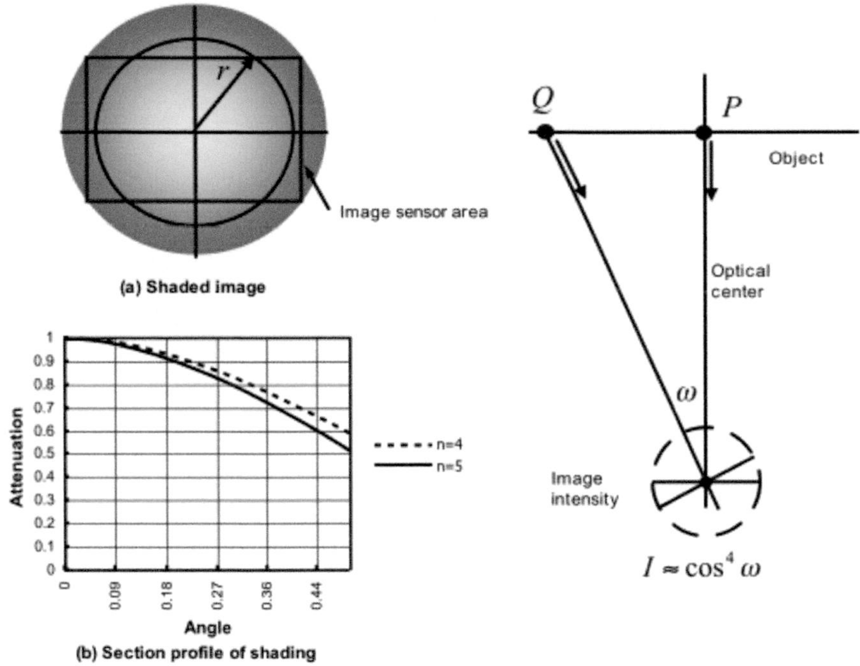

FIGURE 3.29: Lens shading distortion. Figure from [334].

take into account an estimation of image edges so as not to blur them and decrease contrast.

4. Gamma correction should come last in the pipeline because of the non-linear nature of this operation, which makes it difficult to preserve chroma values afterwards.

On the other hand, Weerasinghe et al. [377] propose a pipeline where color correction precedes demosaicking, justifying this choice with an increase in SNR of the output. In fact, in [89] Bazhyna points out that several algorithms for image processing before demosaicking have recently been introduced that deal with digital zooming, denoising, deblurring and sharpening, among other tasks.

Part III

Action

Chapter 4

Compression

4.1 Introduction

If we were to use video data in a direct form, representing each pixel of each frame as a triplet of red, green and blue values expressed with a fixed number of bits, then the amount of data involved would be enormous in every regard: recording, storage, transmission. For instance, recording images of 2880x1620 pixels at 30 frames per second and with 10 bits of color information per channel would require a data rate of approximately 4 gigabits per second. This figure is beyond the capabilities of todays' cameras. A bandwidth of 4 Gb/s is also exceedingly larger than current typical values for domestic DSL connections, which are in the range of a few tens of Mb/s. Finally, storing 90 minutes of such a movie would require a capacity of 2.6 terabytes, equivalent to more than 500 DVDs.

Since cameras are able to record movies that we can watch with good quality by streaming or from a disc or file of just a few gigabytes, it is clear then that the amount of data required to represent the movie has been reduced. The process by which this is achieved is called compression. The goal of compression is not only to meet a bitrate requirement, it must do it while at the same time keeping the image quality above a certain level and using methods of affordable computational complexity [345]. In this chapter we will follow the excellent book by Shi and Sun [345] and present just an overview of the main compression techniques currently used for cinema; the interested reader is referred to that book for details.

4.2 How is compression possible?

Another way of posing the problem of compression would be the following: we want to remove information from the video, but do it in such a way that the viewer doesn't notice.

By their very nature, movies have a lot of temporal redundancy. From one frame to the next, there is usually just a small percentage of the pixels that change their value. We could think of performing compression of a frame by representing only the pixels that change, ommitting those that have the same value as in previous frames. Even when the camera is moving and/or many scene objects are moving, in which case we could argue that most pixels are not constant but changing, a frame can be very well approximated by a version of the previous frame where all objects have been displaced according to their motion, so the temporal redundancy is still there.

Within a frame there is also significant redundancy. This spatial redundancy is clear in homogeneous regions, where each pixel is very similar to all its neighbors, but is also present at image contours (the similarity or redundancy takes place along the contour) and even textured regions (the similarity is restricted to pixels with the same texture pattern). Compression may be achieved then by expressing pixels not by themselves but in terms of their neighbors.

The underlying assumption, both in spatial and temporal redundancy, is that the movies we are talking about are "natural" in a statistical sense, they are not a collection of random dots but a succession of images depicting objects with uniform regions, contours and patterns, and which do not often change abruptly from one frame to the next. This is how we expect movies to be so it makes sense that compression algorithms exploit these properties, but if movies were radically different (e.g. the random dots just mentioned) the compression techniques would also have to be of another kind.

Another key element that makes compression feasible is the fact that human visual perception is different from camera sensing. The same pixel value is perceived easier in some situations than in others, for instance depending on its neighbors, the average value of the image, etc. We may compress, then, by using less data to represent regions that are perceived less. The influence of a stimulus on the detectability of another is called masking. In visual perception there are several types of masking that are exploited for compression purposes:

- *Luminance masking.* Weber's law states that if we have a stimulus I to which is superimposed an increment ΔI, we are able to detect the increment if it is above a certain value $JND(I)$ or "Just Noticeable Difference" which satisfies

$$\frac{JND(I)}{I} = k, \tag{4.1}$$

  where k is a constant for each given sense (hearing, taste, etc.) and for human vision we have one value of k for each of the four types of photoreceptors in the retina [386]. Therefore, if the background of an image is bright, a larger increment is needed to be detected. We may use this in compression when quantizing a signal, assigning more levels for darker values and less levels for brighter regions.

- *Texture masking.* Errors are much more visible in uniform regions than in textured ones. Therefore, quantization levels should adapt to the intensity variations of image regions.

- *Frequency masking.* Errors are more visible under low frequencies. Quantization errors may be masked by adding Gaussian noise *before* quantization: the low frequency noise of the quantization errors is transformed into high frequency noise, and hence made less perceptible, by the quantization procedure. This type of masking is picture independent, unlike texture masking.
- *Temporal masking.* The human visual system takes a while to adapt to abrupt scene changes. During this period it is less sensitive to details, and images may be represented in a coarser way.
- *Color masking.* We are more sensitive to luminance than to chrominance. For a given spatial frequency, the amount of contrast increment needed to detect a difference in luminance is much smaller than the amount needed to detect a change of chrominance in any representation (hue and saturation, blue-yellow and red-green, etc.)

Finally, for compression we may also exploit coding redundancies. This involves estimating the statistics for the symbols involved in the representation of the images, and using less bits for symbols that are more frequent and more bits for symbols with a lower ocurrence probability. For the same image, a better codc will produce a lower average bitrate.

4.2.1 Measuring image quality

One of the requirements of any compression algorithm is to achieve a certain visual quality in the reconstructed images. And measuring visual quality is important when choosing a compression method: if two methods produce the same visual quality we will select the one with the lower bitrate, and if they have the same bitrate we will prefer the one with the higher visual quality.

Ideally, quality measurement should be subective, based on user studies, since what we are ultimately interested in is the perceived quality of the images as they are seen by some "standard" observer. But user studies are complicated and cumbersome, difficult to set up because the visual conditions of the experiment room are paramount, and they are difficult to compare among one another, with results heavily dependent on the image sets and the question asked.

This is why, in practice, objective quality measures are preferred. The most common ones are based on the mean square error (MSE) between the original image f and its reconstruction after compression g:

$$MSE = \frac{1}{M \cdot N} \sum_{x=0}^{M-1} \sum_{y=0}^{N-1} (f(x,y) - g(x,y))^2 \tag{4.2}$$

$$SNR = 10log_{10}\left(\frac{\frac{1}{M \cdot N}\sum_{x=0}^{M-1}\sum_{y=0}^{N-1}(f(x,y))^2}{MSE}\right) \tag{4.3}$$

$$PSNR = 10log_{10}\left(\frac{(max_{x,y}\{f(x,y)\})^2}{MSE}\right). \tag{4.4}$$

These three measures are very easy to compute and by far the ones more used in practice. Higher visual quality implies smaller MSE and larger SNR and $PSNR$ values. The downside is that often these measures are not related to perceived (subjective) quality. Wang and Bovik show in [372] a very telling example where the same image has been modified with different transformations, and while some results are clearly preferrable to others (e.g. changing the mean of the image is better in terms of visual quality than blurring it with a Gaussian kernel), all the transformed versions have the same MSE value; see Figure 4.1.

The field of image quality assessment is very active, see for example the excellent survey by Chandler [117].

4.3 Image compression with JPEG

JPEG stands for Joint Photograhics Experts Group, the name of the committee who worked on the first version of the standard from 1986 to 1992. It is the most popular method for compressing still images, and became a standard in 1992.

JPEG takes as input a 24-bit RGB picture, and first converts it to YCbCr, which is the color space used in TV, DVDs and Blu-rays. By expressing colors not as red, green and blue triplets but in terms of luminance Y and chroma components Cb and Cr, the YCbCr color space allows better perceptual quality than RGB for the same compression rate: by exploiting the color masking property of the human visual system, the Cb and Cr channels can be represented more coarsely (i.e. compressed more) without an impact in the visual quality. Often, color-space conversion is followed by down-sampling of chroma components. The notation is:

- 4:4:4, no subsampling.
- 4:2:2, reduction of the chroma components by a factor of 2 in the horizontal direction.

FIGURE 4.1: Different distortions of the same image. (a) Mean shift. (b) Contrast stretch. (c) Gaussian blur. (d) JPEG compression. They all have approximately the same MSE value with respect to the original. Figure from [372].

– 4:2:0, reduction of the chroma components by a factor of 2 in both the horizontal and vertical directions.

After this step and for the rest of the process the three channels are treated independently, so whenever we mention "the image" we are referring to either of its Y, Cb or Cr channels.

Following color conversion and subsampling, the image is split in 8×8 blocks. If image dimensions are not a multiple of eight the encoder must fill in the remaining area. Using a constant color for fill in creates unpleasant ringing artifacts, so it's better to replicate pixels at the boundary, using for instance mirror symmetry.

Each block is shifted by -127, so the range of values is transformed from

$[0, 255]$ to $[-127, 128]$. This operation reduces the dynamic range in the subsequent processing.

4.3.1 Discrete cosine transform

Next, the DCT (Discrete Cosine Transform) is applied to the block. The DCT is, like the discrete Fourier transform, a linear transform that expresses a signal as a linear combination of basis functions.

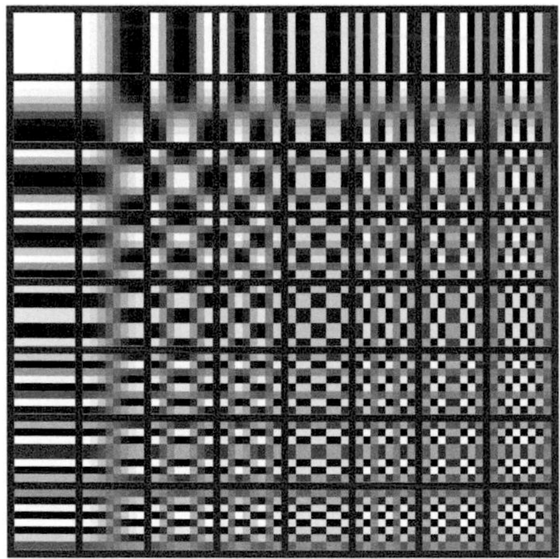

FIGURE 4.2: JPEG's DCT basis functions. Figure from [16].

A very basic property of all linear transforms is that both the signal and its transform have the same number of samples. For example, the DCT of an 8×8 block will consist of 64 transform coefficients, each being the sum of the pixel-by-pixel product of the signal block with one of the basis functions (which is also an 8×8 block). Figure 4.2 shows the 64 basis functions of the DCT: any possible image block can be expressed, with perfect accuracy, as a linear combination of the basis blocks in the figure. Therefore, just by applying the transform we are not achieving any compression; on the contrary, the range of values of the transform coefficients is larger than that of the original values so we would need even more bits to represent the transformed signal than the original one. But the transformed signal is less correlated, and different coefficients correspond to different spatial frequencies. Here JPEG uses the frequency masking property, quantizing more coarsely those coefficients that have a higher spatial frequency. This is achieved by dividing each transform coefficient by a frequency-dependent normalization parameter, and then rounding off the result: as a consequence, many of the quantized coefficients become zero, and compression is achieved by keeping the value and location of non-zero coefficients only. With higher normalization parameters more nor-

Luminance Quantization Table							
16	11	10	16	24	40	51	61
12	12	14	19	26	58	60	55
14	13	16	24	40	57	69	56
14	17	22	29	51	87	80	62
18	22	37	56	68	109	103	77
24	35	55	64	81	104	113	92
49	64	78	87	103	121	120	101
72	92	95	98	112	100	103	99

Chrominance Quantization Table							
17	18	24	47	99	99	99	99
18	21	26	66	99	99	99	99
24	26	56	99	99	99	99	99
47	66	99	99	99	99	99	99
99	99	99	99	99	99	99	99
99	99	99	99	99	99	99	99
99	99	99	99	99	99	99	99
99	99	99	99	99	99	99	99

FIGURE 4.3: Two examples of quantization tables. Figure from [345].

malized coefficients will be rounded to zero and more compression will be obtained. The 64 normalization parameters constitute the 8×8 quantization table Q, and two examples of such a table are shown in Figure 4.3.

Q tables are derived experimentally, through user studies, by presenting obervers with gratings of varying spatial frequency and measuring the discrimination thresholds of the subjects [309]. Notice how the elements of the Q table for luminance are in general quite smaller than those of the Q table for chrominance. This is a consequence of the color masking property of human vision, and implies that at the same visual quality we can compress the chroma information more, since we are dividing the chroma coefficients by larger normalization values and therefore more of them will be rounded off to zero. Also notice how in both tables the values increase as we move to the right and to the bottom, i.e. as we increase spatial frequency. This is a consequence of the freqency masking property of the visual system.

4.3.2 Run-length and entropy coding

Further savings are obtained by using a zig-zag scan and run-length coding (RLC). The zig-zag scan is just an ordered list of the transform coefficients going from lowest to highest spatial frequency; see Figure 4.4. With RLC, each non-zero coefficient is represented with two numbers, one is the coefficient value and the other is the number of zero-coefficients immediately before it in the zig-zag scan. Finally, numbers in RLC are binary-coded with a Huffman code, using probability tables derived experimentally and which appear in an annex of the JPEG standard [115].

The RLC procedure applies to all transform coefficients except the one with spatial frequency $(0, 0)$. This is called the DC coefficient (after "direct current," using the term from electrical engineering denoting current with constant amplitude), and corresponds to the mean average of the signal block. An error in the DC coefficient is highly visible, as a change in average with respect to the neighboring blocks, and therefore this coefficient is compressed without loss: no normalization and rounding, just a simple differential coding, i.e. what's coded is the difference between the DC of the current block and the DC of the previous one (the DC of the first block is coded as it is).

FIGURE 4.4: JPEG's zig-zag scan of DCT coefficients. Figure from [18].

4.3.3 Reconstruction and regulation of the amount of compression

To reconstruct the image, all but one of the above steps are inverted: from the Huffman binary words the normalized and zero coefficients are extracted, then the normalized coefficients are multiplied by the quantization table values, the DC is recovered, then the inverse DCT is applied, a value of 127 is added to every element, and finally the colorspace is converted from $YCbCr$ to RGB. The one step that can't be inverted is the quantization performed by rounding off the normalized coefficients, and this is the step that allows for compression. The amount and quality of the compression are regulated by the quantization table Q: for more compression, and less quality, Q is multiplied by a factor (e.g. determined when the user selects a given quality for the reconstructed image). Figure 4.5 shows different JPEG outputs for the same input image, varying the quality of the compression. The uncompressed original has a size of $3.9MB$. Notice how, at 95% quality, we achieve virtually no quality loss with respect to the original, but a compression ratio of approximately $8:1$. As the quality is reduced the factor multiplying the quantization tables increases, and the final result becomes poorer. At low qualities visual artifacts become very visible and take the shape of abrupt changes in luminance and color between neighboring blocks, due to the JPEG processing that is performed independently for each block. This is why compression artifacts are often called block or blocking artifacts. A larger block size would make these artifacts less objectionable and the compression rate would also be increased; the downside would be an increased computational cost and a decreased resiliency to errors, since an error in a transform coefficient affects the reconstruction of the whole block. The JPEG block size of 8×8 was arrived at as a compromise between all these aspects.

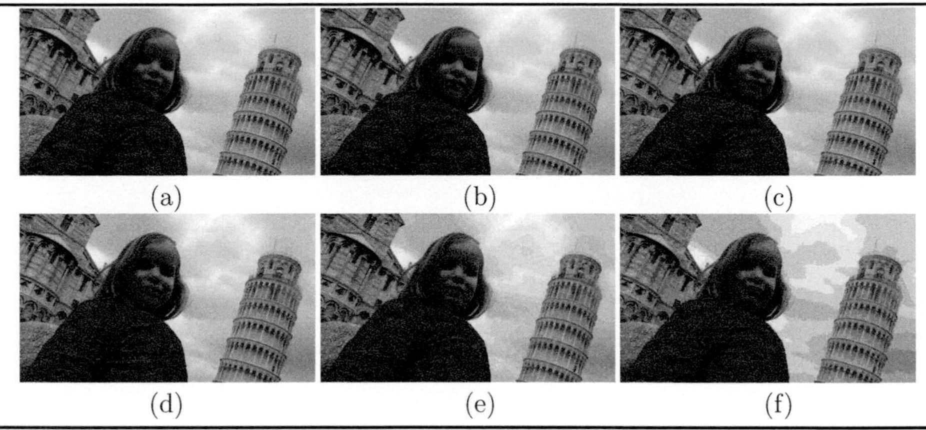

FIGURE 4.5: Different JPEG outputs varying the quality of the compression. The uncompressed original has a size of $3.9MB$. (a) 95% quality, $507KB$ size. (b) 75%, $138KB$. (c) 50%, $90KB$. (d) 25%, $58KB$ size. (e) 10%, $33KB$. (f) 5%, $22KB$.

4.4 Image compression with JPEG2000

The JPEG2000 image compression standard uses the discrete wavelet transform (DWT) instead of the DCT. The basic idea is the same: to transform the signal onto another representation so that the transform coefficients are de-correlated, then perform compression by thresholding or rounding some of the coefficients to zero. Possibly the main difference between the DWT and the DCT is that the DWT is not applied on blocks but on the whole image, at different scales. In a 1D DWT the signal is filtered with two filters, a lowpass and a highpass, both outputs are then downsampled by a factor of two, and this process (filtering and downsampling) is iterated for the lowpass subband, the downsampled output of the lowpass filter, which is itself filtered with a lowpass and a highpass filter, downsampled, etc. Iterations are performed an estipulated number K of times; K is called the number of levels of the wavelet transform. For an image, a 2D signal, the DWT is computed by applying a 1D DWT on the rows, then on the columns. Figure 4.6 shows an image, its 3-level wavelet transform and the naming scheme of the wavelet subbands. The letter L stands for lowpass, H for highpass, a pair LH means lowpass filtering of the rows and highpass filtering of the columns (and so on for all other pairs HL, LL, HH), and the subindex of the pair indicates the level, e.g. LH_2 corresponds to level two.

After wavelet coefficient thresholding, the location of the zero coefficients

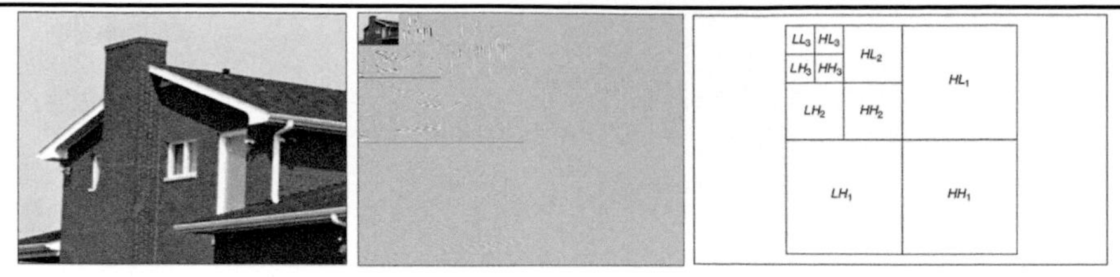

FIGURE 4.6: Left: original image. Middle: subbands of its 3-level wavelet transform. Right: naming scheme of the wavelet subbands. Figure from [365].

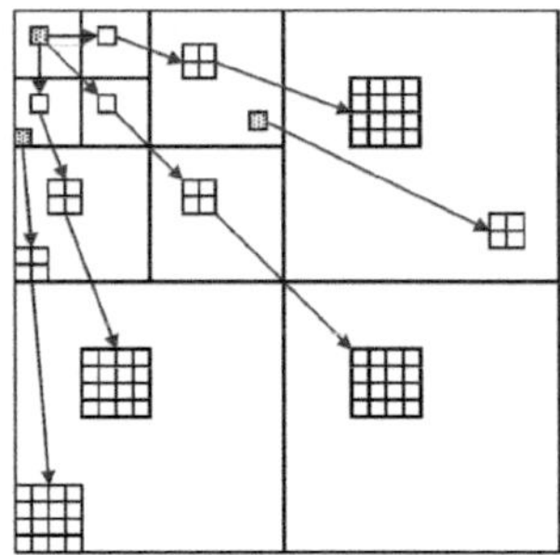

FIGURE 4.7: Example zero trees with roots indicated by shading. Figure from [365].

can be coded very efficiently using a structure called "zero tree" (see Figure 4.7) which exploits the fact that the zeroes tend to be located at the same spatial location at different scales: *"if a wavelet coefficient at a coarse scale is insignificant with respect to a given threshold T, then all wavelet coefficients of the same orientation in the same spatial location at finer scales are likely to be insignificant with respect to T"* [365].

JPEG2000 has less objectionable artifacts than JPEG at high compression ratios, as Figure 4.8 shows.

4.5 Video compression with MPEG-1, MPEG-2, MPEG-4 AVC (H.264)

In order to compress video, we could just apply JPEG or JPEG2000 compression independently to each of the frames of the video sequence. This is

FIGURE 4.8: Left: original image. Middle: JPEG compression, 100:1 ratio. Right: JPEG2000 compression, 100:1 ratio. Images from [151].

in fact the approach used in practice by many consumer camera models that record in MJPG (motion-JPEG), and by some high-end digital cinema cameras that compress each frame using the wavelet transform. But if we take into account the large temporal redundancy present in most video sequences, we can achieve significantly higher compression rates. Instead of coding a whole frame, we can code only the differences with respect to the preceding frame: as we argued in section 4.2, by their very same nature movie frames don't usually change much from one to the next. A frame can be very well approximated by a version of the previous frame where all objects have been displaced according to their motion, and this technique, called *motion compensation*, is at the core of all the MPEG video compression standards. MPEG stands for Moving Pictures Experts Group, the committee that developed:

- MPEG-1 (1991): for multimedia CD-ROM applications, coding up to 1.5Mb/s. Features: frame-based random access; fast forward/backward searches through compressed bitstreams; reverse playback; editability of compressed stream.

- MPEG-2 (1994): for standard definition TV quality (2-15Mb/s). Used in digital TV, some HDTV, all DVDs, some Blu-Rays. It is a superset of MPEG-1, so every MPEG-2 decoder is also capable of decoding MPEG-1 streams. MPEG-2 enhancements with respect to MPEG-1 are mainly coding performance improvements related to the suport of interlaced material, downloadable (optimized) quantization tables, alternative scan orders (other than the zig-zag scan) for encoding transform coefficients, and also the possibility of scalability, which allows a subset of the video stream to be decoded into meaningful video thus increasing error resiliency (at the cost of an increased overhead).

- MPEG-4 AVC/H.264 (2003): for HDTV, Blu-Ray discs, streaming Internet sources. Provides 50% bitrate savings with respect to MPEG-2, at an equivalent perceptual quality. Coding efficiency has been improved in two ways: by increasing the accuracy of the prediction of the picture to be encoded (by using variable block sizes, multiple image references, directional spatial prediction -edge extrapolation-, deblocking filters not as post-processing but already in the coding stage, etc.) and by using improved methods of transform and entropy coding (an integer version of the DCT, binary arithmetic coding, etc.)

For each frame in MPEG-1 and MPEG-2, and each frame-slice in H.264[1] there is the possibility of using intra-frame coding or predictive coding. When a frame is compressed independently from the others, treating it as a standalone image, then it is said to be of type I for "intra-coded," coded by intra-frame techniques. In MPEG-1 and MPEG-2 this is very similar to JPEG coding. Frames encoding (motion-compensated) differences with another frame are called type P for "predictive-coded," or B for "bi-directionally predictive-coded." The image sequence is partitioned into groups of pictures, GOPs, which usually last from 12 to 15 pictures. Each GOP is independent from the others: all P and B frames in a GOP code motion-compensated differences with other I and P frames in the same GOP. Frames of type I and P are thus called anchors. An error in the first I frame of a GOP may affect all subsequent P and B frames in the GOP, while an error in a P frame will only affect some P and B frames and an error in a B frame does not propagate. Figure 4.9 shows an example of a GOP in MPEG-1.

Since one requisite of all MPEG video compression standards is to allow random access to any frame in the sequence, I frames must be used from time to time; otherwise, the time required to reconstruct a frame in the middle of the sequence would be much larger than the time required to reconstruct a frame at the beginning, just because the predictive coding implies reconstructing all preceding frames up to the previous I frame. In MPEG-1, GOPs start with an I frame and this is usually the only I frame of the GOP.

A P frame is coded with one-directional motion compensation. Let R be the reference frame of this frame P (R itself may be a P or an I frame, as we mentioned). The P frame is split into non-overlapping blocks, and for each block b_P in P its most similar block b_R in R is found (through some block-matching technique): the diference in location among these matching blocks is a motion vector $\hat{v}_b$ that is associated to the block b_P. It is called a motion vector because it describes the motion undertaken by b_R so as to arrive to the position of b_P. But coding moving objects as same-sized blocks and representing motion only as traslation is often a rather crude approximation. Therefore, if we were to reconstruct P just by placing at each location b_P the block b_R indicated by the motion vector $\hat{v}_b$, the result could be quite poor. This is why prediction coding involves not only the motion vectors but also the prediction error, the difference $b_P - b_R$ between the original b_P and the

[1] A slice is a region of the frame defined as a set of non-overlapping 16×16 macroblocks, and it has arbitrary shape, size and topology.

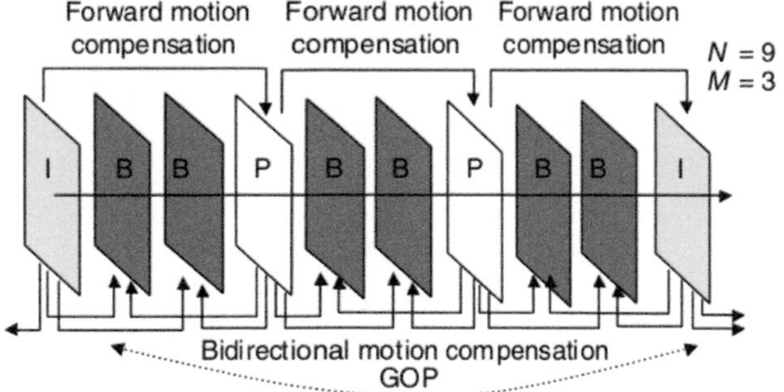

FIGURE 4.9: Example of an MPEG-1 GOP. The GOP length is $N = 9$, the distance between two anchors is $M = 3$. Figure from [345].

predicted value b_R. Motion vectors are coded with difference coding (DPCM, no rounding) and the prediction error is treated as an image and coded with the DCT (with different quantization tables than those used in JPEG, since the statistics of a prediction error image are not the same as the statistics of a natural image).

A B frame is coded using the same approach, i.e. representing motion vectors and prediction error, but the references are two (I or P) frames instead of one. B frames require larger data buffers (to store I and P frames) and cause delay, which is why they are sometimes avoided. On the other hand, they are useful when the background behind an object is revealed over several frames, or in fading transitions. They can also help control bitrate: a very low bitrate B frame may be inserted when needed and this is not problematic because no frame depends on it.

4.6 In-camera compression

We review here some concepts on compression introduced in the chapter on cameras. Virtually all cameras perform compression and only a few models, like the ARRI Alexa, allow for uncompressed recording but this must be performed off-camera and requires special recording equipment.

The compression may be lossless, in which images can be recovered without any information loss, or lossy, with information loss increasing with the compression rate.

Lossless compression is based on Huffman coding and run-length encoding and is normally used for raw data, as in the CinemaDNG format, based on the TIFF/EP format with which raw images are recorded in several models of still cameras.

For lossy compression there are many alternatives. For starters, the color coding can be in RGB or YCbCr, and in YCbCr there are different possibilities of color subsampling.

Next, there is the possibility of using intra-frame coding or predictive coding. In several formats such as ProRes, DNxHD, and REDCODE, all frames are of type I, intra-coded. As we said earlier, a GOP always starts with an I frame, so if a codec only uses one-frame GOPs it is in fact coding all frames with intra-coding and all frames are independent. Videos in which GOPs have more than one picture are called Long-GOP. Very popular formats like MPEG-1, MPEG-2 and MPEG-4/AVC H.264 use predictive coding, i.e. their frames may be of type I, P or B. If a video in one of these formats uses Long-GOP it can't be edited directly due to the presence of prediction-coded frames P or B; it must be converted to a format suitable for editing in a process called "rushes upgrade," going for instance from H.264 to ProRes 422 [143].

Finally, there's the actual lossy compression technique used, which can be based either on the DCT or on the wavelet transform. The DCT is the basis for MPEG-1, MPEG-2, MPEG-4/AVC H.264, DNxHD and ProRes. The wavelet transform is the basis for REDCODE and DCI $X'Y'Z'$ [34], both of which use only I frames coded with variants of JPEG2000.

Chapter 5

Denoising

5.1 Introduction

Noise is always present in images, regardless of the way they have been acquired. Movies shot in film have an intrinsic film grain, digital movies have noise due to the acquisition process (Chapter 3) and the subsequent compression (Chapter 4) and transmission. Denoising is therefore a very important problem in image processing, with a vast literature on it and with early works dating back to the 1960s.

Classic denoising techniques were mostly based in one of these two approaches: modification of transform coefficients (using the Fourier transform, the DCT, some form of wavelet, etc.), or averaging image values (in a neighborhood, along contours, with similar but possibly distant pixels, etc.) Both types of approaches yielded results that were modest in terms of PSNR and also in visual quality, with frequent problems such as oversmoothing, staircase effects, or ringing artefacts.

In 2005, two groundbreaking works independently proposed essentially the same idea: comparing image neighborhoods (patches) in order to denoise single pixels. These so-called non-local approaches produced results that were shockingly superior to the state of the art at the time, so much so that from then on virtually all image denoising algorithms have been patch-based. Actually, the increase in quality of the denoising algorithms in the past few years has been so dramatic that several recent works have asked if there is still room for improvement in denoising and how far we are from the optimum: the answer appears to be that we're close.

In this chapter we'll start by presenting the basic ideas of classical image denoising techniques, followed by the introduction of non-local approaches and the main works that have come after them. Then we'll comment on optimality bounds for denoising and finally we'll introduce a recent work where we propose a novel framework: denoising an image by denoising its curvature instead.

5.2 Classic denoising ideas

In their outstanding survey on denoising, Lebrun et al. [246] explain that classic or early denoising techniques stemmed from the fact that natural images are band-limited, while noise isn't. This led, on one hand, to approaches where the image is transformed and the high-frequency components, assumed to correspond just to the additive noise and not to the band-limited image signal, are set to zero. It also led to approaches where each pixel is replaced by an average of its neighbors, because if the image is band-limited then pixel values must be spatially correlated.

5.2.1 Modification of transform coefficients

These techniques set to zero or just atenuate the transform coefficients that are supposed to correspond to the noise. For instance, using the Fourier transform and a Gaussian function for atenuating the coefficients, we get denoising by convolution with a Gaussian kernel (we also get a blurry result, of course). As the Fourier transform is global it may not be capable of preserving local details so local transforms, like the DCT or the wavelet transform, are preferred. We saw in Chapter 4 how transforms, by themselves, do not reduce the amount of data (actually it takes more bits to represent the transform coefficients than the image values); rather, image compression is achieved by setting to zero small coefficients that statistically are not very frequent in (clean) natural images. This is the same principle of image denoising at work in compression, therefore we can see that images that have been compressed with a lossy technique have also been partially denoised: this is the sort of denoising performed in-camera, where it's called *coring*, as we mentioned in Chapter 3. Yaroslavsky [390] uses a sliding window in which he computes the DCT and atenuates its coefficients. Donoho [145] computes a wavelet transform, sets to zero all coefficients below a certain threshold and atenuates the rest according to their magnitude.

5.2.2 Averaging image values

These techniques replace the value at each pixel by a weighted average of the values of its neighbors. If we consider neighbors in all directions and weights given by a Gaussian kernel centered in the pixel, we get denoising by Gaussian blurring. Averaging values not in all directions but in the direction of image contours improves the result, and this is the route taken by the methods of anisotropic diffusion of Perona and Malik [308], total variation (TV) denoising of Rudin, Osher and Fatemi [330], and more recently by the denoising of normals method of Lysaker et al. [260] and the Bregman iterations approach of Osher et al. [297]. Another route is to consider neighbors in all directions but only if their values are similar, possibly weighting them according to their spatial distance, as proposed by Yaroslavsky in his neighborhood filter [389], Smith and Brady in their SUSAN filter [349] and Tomasi and Manduchi with their bilateral filter [359].

5.3 Non-local approaches

At the Computer Vision and Pattern Recognition conference of 2005 there were two works that completely changed the image denoising landscape and which independently proposed an essentially very similar idea, based on comparing image neighborhoods (patches). These were the UINTA algorithm of Awate and Whitaker [79] and the non-local means algorithm of Buades et al. [109].

The UINTA work starts by pointing out that each neighborhood in an image can be regarded as an observation from the probability density function (PDF) of the image. Instead of imposing some ad-hoc, generic parametric PDF model, as it was usually done, Awate and Whitaker estimate the PDF from the image itself, by comparing image patches. Noisy images have a less regular PDF, with a higher entropy, and therefore the UINTA algorithm performs denoising by iteratively reducing the entropy associated with the PDF.

In the non-local means (NLM) algorithm, the value at each pixel is replaced by a weighted average of all pixels in the image, weighting them according to the similarity of their patch neighborhoods. Figure 5.1 depicts a schematic for this method. Pixel p is replaced by a weighted average of all other pixels, and the weights depend on the similarity of the neighborhoods: q_1 and q_2 will have large weights, whereas the weight for q_3 will be small, despite the fact that the pixel values of p and q_3 might actually be similar, or that q_3 could be spatially closer to p than q_1 or q_2.

NLM is a one-shot algorithm, although it can be iterated: the iteration of NLM is esentially equivalent to UINTA. NLM can be seen as an extension of the neighborhood filter of Yaroslavsky [389] where now whole image patches and not just single pixels are compared. It can also be seen as an adaptation to denoising of the non-parametric estimation principles of the (also seminal) texture synthesis work of Efros and Leung [153] (mentioned in Chapter 14 as a groundbrealing inpainting method), as explained in [246].

Both UINTA and NLM produced results of a quality dramatically superior to what was available at the time, as we can see in Figures 5.2 and 5.3. After these works, virtually all image denoising algorithms and definitely all the state of the art methods have been patch-based. Of the two works, NLM is the one that has become more popular and this is why all subsequent denoising works based on patches are deemed "non-local."

FIGURE 5.1: Schematic of NLM algorithm of Buades et al. [109]. Pixel p is replaced by a weighted average of all other pixels, and the weights depend on the similarity of the neighborhoods: q_1 and q_2 will have large weights, whereas the weight for q_3 will be small. Figure from [109].

5.4 An example of a non-local movie denoising algorithm

In this section we present in some detail a non-local movie denoising algorithm, the average of warped lines (AWL) of Bertalmío, Caselles and Pardo [93], reproducing excerpts and figures from that paper.

The input to the AWL algorithm is a video sequence of T images (*frames*) I^t. We partition the domain of each frame I^t into a disjoint set of lines $\mathcal{L}_i^t$, which for simplicity we take as the horizontal lines $\mathcal{L}_i^t = \{(i,j,t) : 1 \leq j \leq N\}$ (row i in frame t). For each line $\mathcal{L}_i^t$ we consider the family of its neighboring ones, both spatial (nearby lines in the same frame) and temporal (nearby lines from other frames). Let us call this set of neighboring lines $\mathcal{NL}_i^t$.

We take each line $\mathcal{L}_m^s \in \mathcal{NL}_i^t$ and *warp* it so as to make it match with $\mathcal{L}_i^t$; see Figure 5.4. The denoised version $\mathcal{D}_i^t$ of I on the original line $\mathcal{L}_i^t$ can be obtained by performing an average (mean, weighted mean, or median) of the warped lines. The same is done for every frame t and line $\mathcal{L}_i^t$ in the video sequence, producing the denoised image $\mathcal{D}(i,j,t)$. This idea is illustrated in Figure 5.5.

We could think in more general terms and apply this warping and averaging procedure to $2D$ regions instead of lines. But we choose the regions $\mathcal{L}_i^t$ to be scan-lines, that is, horizontal lines spanning the whole image from the first to the last column. The reason for this choice is that with $1D$ regions the warping operation reduces to the problem of dense matching in stereo applications, for which there is extensive literature (see for instance [132] and references therein) and, more importantly, we can use dynamic programming to compute a warping between lines that satisfies the *ordering constraint* (that states that if pixel P is before pixel Q in line L, then their correspondent matches in line L', pixels P' and Q', will also satisfy that P' comes before Q') and a *uniqueness*

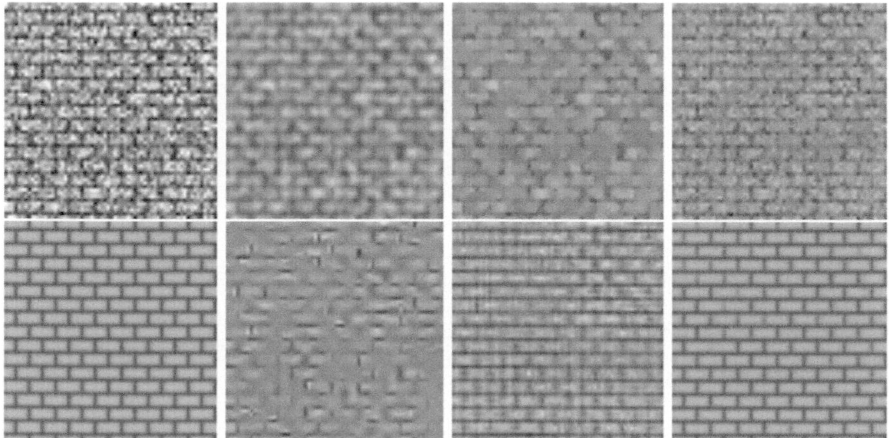

FIGURE 5.2: Comparison of NLM with typical denoising algorithms used at the time (2005). From left to right and from top to bottom: noisy image, Gauss filtering, total variation, neighborhood filter, Wiener filter (ideal), wavelet thresholding, DCT Wiener filter (empirical), NLM. Figure from [110].

constraint when there are no occlusions (which states that any pixel P in line L will only match one pixel P' in line L').

Figure 5.6 shows, on the top row, two 100×100 images that are details of two consecutive frames of the film "The Testament of Dr. Mabuse." On the bottom row of the figure we see the profiles of scanline number 50 in each image. Figure 5.7 shows, on the left, the central diagonal band of the 100×100 cost matrix $C(j,l)$ for matching those lines. In this figure, a light grey value corresponds to a high cost, and a dark grey value corresponds to a low cost. Therefore, matching these lines amounts to finding the path in this matrix (from the top left to the bottom right) which has a minimum accumulated cost. In general, finding the solution to this optimization problem has a complexity of $O(N^3)$, where N is the number of pixels of each line. Introducing the aforementioned constraints of uniqueness and ordering, the complexity reduces to $O(N^2)$. As we see (Figure 5.7, right) the optimal path does not deviate substantially from the diagonal, allowing us to compute just a diagonal band instead of the whole matrix. Restricting the optimal path to such a band reduces the complexity to $O(DN)$ where D is the width of the band around the diagonal of the matrix $C(j,l)$. Since in practice a small value of D (relative to the image size) suffices, we have that the complexity of finding a warping between two lines is $O(N)$. This implies that total complexity of the denoising procedure is $O(M \times N)$ for an image of size $M \times N$, since we are performing a number of warping/matching operations that is independent of the size of the image.

FIGURE 5.3: Comparison of UINTA with a state-of-the-art denoising algorithm at the time (2005). From left to right: clean image, detail, noisy image detail, result with UINTA, result with the wavelet denoising method of Portilla et al. [312]. Figure from [80].

FIGURE 5.4: A non-decreasing warping between two horizontal lines. Matches are shown as overlayed white segments. Intervals without depicted connections are mapped to a single point of the other line. Figure from [93].

5.4.1 Experiments

Figure 5.9 shows on the top left a frame from a film that is heavily corrupted with noise. This noise is signal-dependent and can *not* be modeled as additive. On the top right we see the result of applying Field of Experts (FoE) denoising [275] to this image. On the middle left we see the result of the NL-Means denoising algorithm. On the middle right and bottom row we see the results obtained with our Average of Warped Lines (AWL) technique using mean, weighted and median averages. Figure 5.10 shows enlarged details of these images. The FoE denoising result is a bit over-smoothed: notice the crystal jar in the background, the square tiles, the writing on the sheet of paper on the desk. Also, this method seems to enhance impulse noise or some dirt and dust defects: notice this effect on the face of the actor, the desk, near the legs, etc. Concerning the NL-Means result, it looks clean and sharp, but some parts have been oversmoothed, like the above mentioned square tiles behind the desk or the writing on the paper sheet on the desk. These structures are better respected with AWL. On the other hand, there is another distracting visual artifact present in the NL-Means result which is only noticeable by watching the denoised video. While one frame at a time the images denoised with NL-Means look perfectly reasonable, when we see them *in motion* at a rate of $24fps$ we clearly perceive an oscillation in the boundary of the objects in the scene. In some cases this artifact really catches one's eye, like when the lamp in this scene, which we know to be rigid and static, slightly deforms and

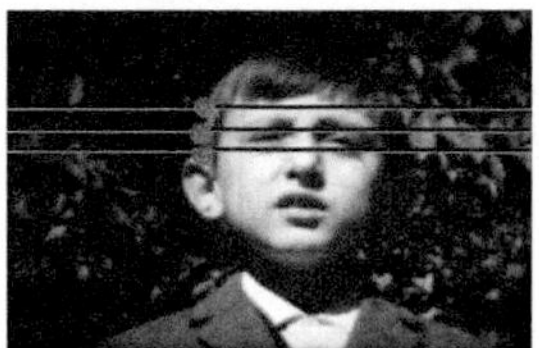 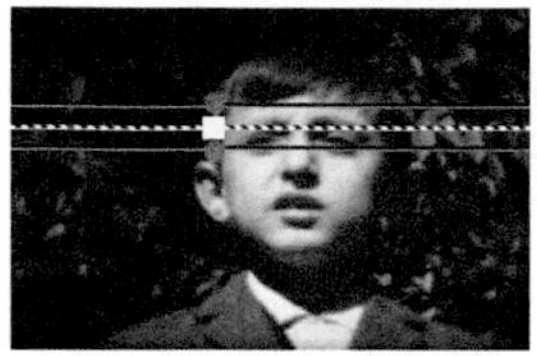 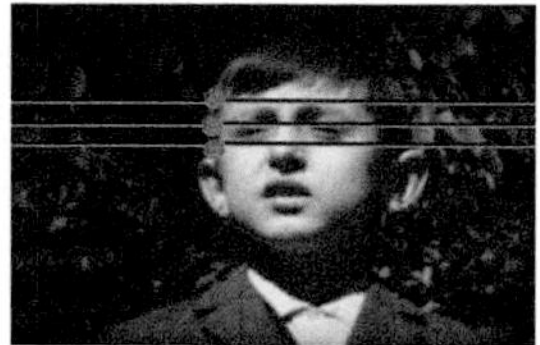

FIGURE 5.5: Basic formulation of AWL algorithm. The striped line in the middle frame is warped to eight other lines, shown in solid red color. For a particular pixel in the striped line (shown as a white square) we show its matches in the other lines (as green circles). AWL performs denoising by replacing the gray value at the white square pixel by an average of the gray values of the green circle pixels. The lines chosen for warping are not adjacent, but this is only for clarity. Figure from [93].

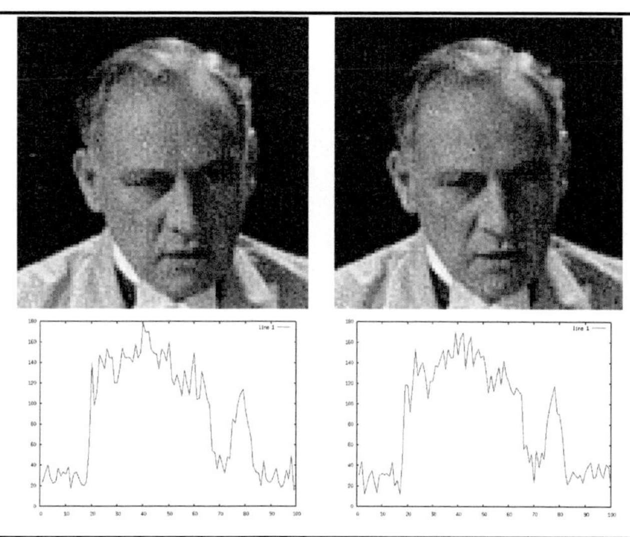

FIGURE 5.6: Top row: 100×100 details of two consecutive frames of the film "The Testament of Dr. Mabuse." Bottom row: corresponding profiles of scanlines number 50 in each image. Figure from [93].

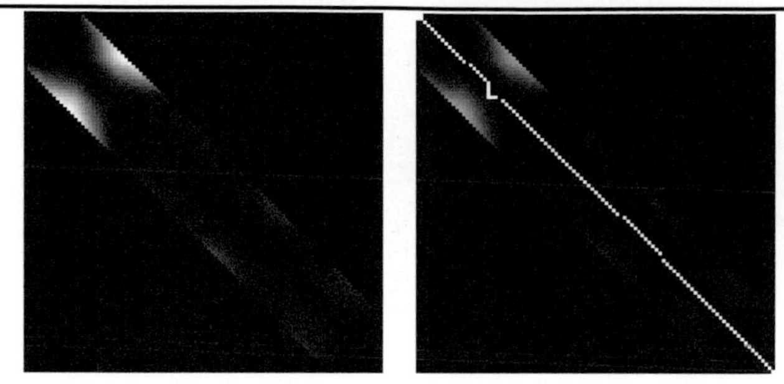

FIGURE 5.7: Left: cost function for matching line number 50 in the images shown in Figure 5.6. Right: same cost function with optimal path superimposed. Figure from [93].

seems to be bending from one frame to the next. Figure 5.8 points this out by showing a detail (corresponding to the lamp just mentioned) of the difference between the original and denoised images both for NL-Means and AWL (with median average) for frame 66 in the sequence. Observe that the boundary of the lamp is clearly visible for the NL-Means case. This means that the process of NL-Means denoising has introduced an error of order $0.5 - 1.0$ pixels in the structures present in the image, and while this error is acceptable for still images it is not for moving images, where it immediately calls the viewer's attention.

The cases of oversmoothing of the NL-Means algorithm may be explained by the fact that it uses all pixels of the image (or a block around the given pixel) in the averaging process. Although each pixel is weighted by its similarity to the pixel to be denoised, if the number of outlier pixels is greater than the number of correct similar pixels the result will be biased. This is the case along edges where a difference of one pixel normal to the edge may produce a small distance and therefore a large weight.

5.5 New trends and optimal denoising

Of the many denoising algorithms that have followed the seminal non-local works we will now mention three, apparently quite dissimilar, methods, each one a representative of a certain approach to the denoising problem. But as Lebrun et al. point out in [246], these methods are almost equivalent: they all estimate local models and denoise a patch by finding its likeliest

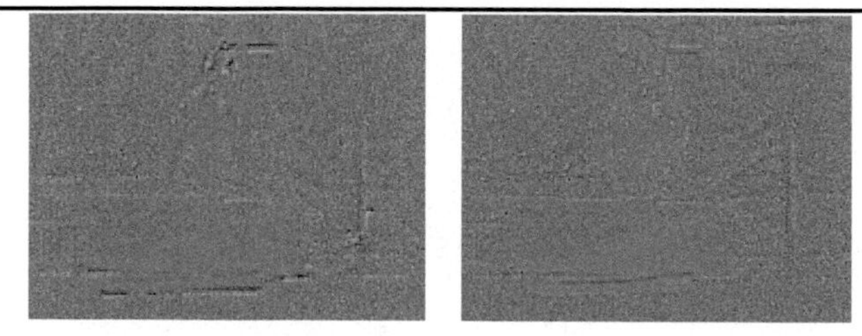

FIGURE 5.8: Detail of difference between original and denoised, for frame 66 in the Mabuse sequence. Left: difference between original and denoised with NL-Means. Right: difference between original and denoised with AWL (median average). Figure from [93].

interpretation. Furthermore, these methods have very similar performance, which recent works suggest is very close to the optimum that can be achieved.

Elad and Aharon [154] propose a denoising method based in sparse representation and the K-SVD algorithm. An image dictionary of patches is built, and each patch of the noisy image is expressed as a linear combination from the dictionary patches, plus an error. A triple-nested optimization procedure looks for the smallest dictionary where the noisy image patches are a combination of the smallest number of dictionary patches and produce the smallest errors. The output of the optimization procedure is the denoised image.

Dabov et al. [137] propose the very popular BM3D (Block Matching 3D) algorithm, arguably the best type of denoising technique available. BM3D combines NLM with transform coefficient thresholding: each noisy patch is grouped into a 3D block with other similar patches, this block is transformed with a linear 3D transform (with a fixed basis), and the coefficients are hard-thresholded. All patches in the block are denoised simultaneously, and after the inverse transform an aggregation step performs a weighted average of all image patches that are overlapping. A subsequent stage refines the output by performing again 3D grouping and 3D Wiener filtering.

Levin and Nadler [249] propose a variation on NLM where instead of considering only the patches present in the image they represent the distribution of natural images using a huge set of 10^{10} patches. This is a very impractical approach, taking about a week of computation on a 100-CPU cluster to denoise 2,000 patches. But the main point of the article is to take this set of 10^{10} noise-free patches as an unbiased estimate of noise-free natural image estatistics, and use it to estimate an upper bound on the PSNR that can be achieved with a patch-based method. They conclude that for small windows, state-of-the-art denoising algorithms such as BM3D are approaching optimality and cannot be further improved beyond fractions of a dB (actually, Lebrun et al. [246] comment that BM3D belongs to a less-restrictive class of algorithms than that used by Levin and Nadler in their study, therefore the

FIGURE 5.9: Original image (top left), denoised by FoE denoising (top right), denoised by NL-Means (middle left), denoised by AWL with mean average (middle right), denoised by AWL with weighted average (bottom left), denoised by AWL with median average (bottom right). Figure from [93].

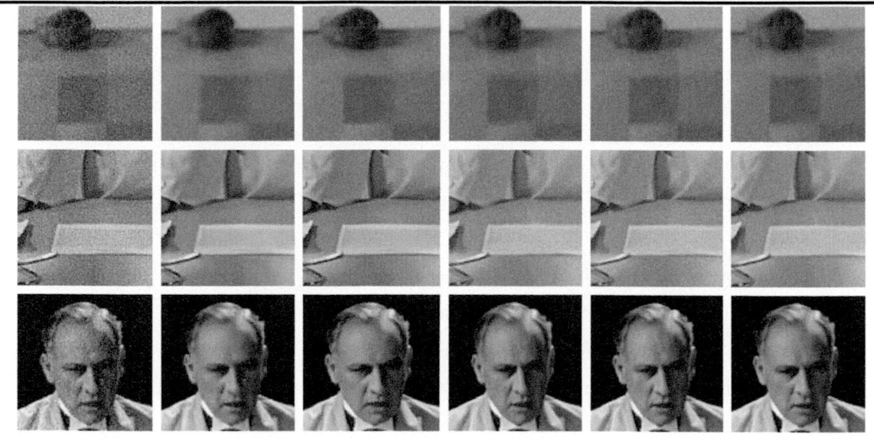

FIGURE 5.10: Details of Figure 5.9. Columns from left to right: Original image, denoised by FoE denoising, denoised by NL-Means, denoised by AWL with mean average, denoised by AWL with weighted average, denoised by AWL with median average. Figure from [93].

actual optimal bound may be larger). A very similar view, that denoising algorithms have little room for improvement, has been put forth by Chaterjee and Milanfar in a work suggestively titled "Is denoising dead?" [118].

5.6 Denoising an image by denoising its curvature image

In this final section we present a novel framework for denoising, by Bertalmío and Levine [100], where we argue that when an image is corrupted by additive noise, its curvature image is less affected by it, i.e. the PSNR of the curvature image is larger. We speculate that, given a denoising method, we may obtain better results by applying it to the curvature image and then reconstructing from it a clean image, rather than denoising the original image directly. Numerical experiments confirm this for several PDE-based and patch-based denoising algorithms. The following contains excerpts and figures from [100].

5.6.1 Image noise vs. curvature noise

We would like to answer the following question: when we add noise of standard deviation σ to an image, what happens to its curvature image? Is it altered in the same way?

Let's consider a grayscale image I, the result of corrupting an image a with additive noise n of zero mean and standard deviation σ,

$$I = a + n. \tag{5.1}$$

We will denote the curvature image of I by $\kappa(I) = \nabla \cdot \left(\frac{\nabla I}{|\nabla I|}\right)$. For each pixel x, $\kappa(I)(x)$ is the value of the curvature of the level line of I passing through x. Figure 5.11(a) shows the standard lena image a and Figure 5.11(b) its corresponding curvature image $\kappa(a)$; in Figures 5.11(c) and (d) we see I and $\kappa(I)$, where I has been obtained by adding Gaussian noise of $\sigma = 25$ to a. Notice that it's nearly impossible to tell the curvature images apart because they look mostly gray, which shows that their values lie mainly close to zero (which corresponds to the middle-gray value in these pictures). We have performed a non-linear scaling in Figure 5.12 in order to highlight the differences, and now some structures of the grayscale images become apparent, such as the outline of her face and the textures in her hat. However, when treating the curvature images as images in the usual way, they appear less noisy than the images that originated them; that is, the difference in noise between a and I is much more striking than that between $\kappa(a)$ and $\kappa(I)$.

This last observation is corroborated in Figure 5.13 which shows, for Gaussian noise and different values of σ, the *noise* histograms of I and $\kappa(I)$, i.e. the histograms of $I - a$ and of $\kappa(I) - \kappa(a)$. We can see that, while the noise in I is $N(0, \sigma^2)$ as expected, the curvature image is corrupted by noise that, if we model as additive, has a distribution resembling the Laplace distribution, with standard deviation smaller than σ. Consistently, in terms of Peak Signal to Noise Ratio (PSNR) the curvature image is better (higher PSNR, less noisy) than I, as is noted in the figure plots.

Another important observation is the following. All geometric information of an image is contained in its curvature, so we can fully recover the former if having the latter, up to a change in contrast. This notion was introduced as early as 1954 by Attneave [77], as Ciomaga et al. point out in a recent paper [129]. Thus, if we have the clean curvature $\kappa(a)$ and the given noisy data I (which should have the same average contrast along level lines as a), then we should be able to recover the clean image a almost perfectly. One such approach for doing this could be to solve for the steady state of

$$u_t = \kappa(u) - \kappa(a) + \lambda(I - u), \; u(0, \cdot) = I \tag{5.2}$$

where $\lambda > 0$ is a Lagrange multiplier that depends on the noise level. As $t \to \infty$, one can expect that $u(t, \cdot)$ reaches a steady state $\hat{u}$ (this is discussed further in Section 5.6.3). In this case, $\kappa(\hat{u})$ should be close to $\kappa(a)$ and the average value of $\hat{u}$ (along each level line) stays close to that of a.

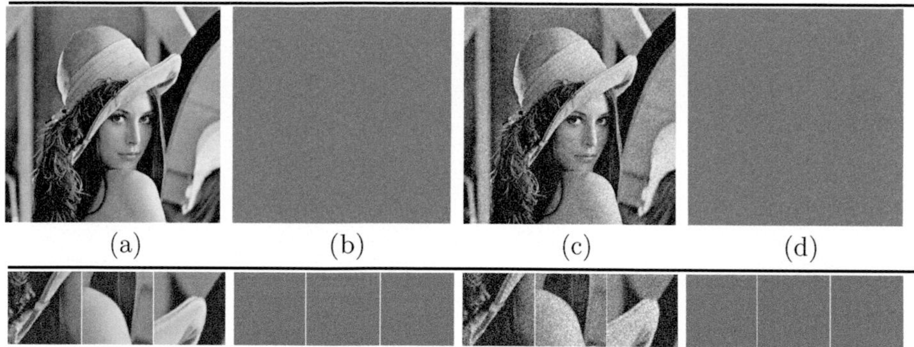

FIGURE 5.11: (a), (b): image and its curvature. (c), (d): after adding noise.

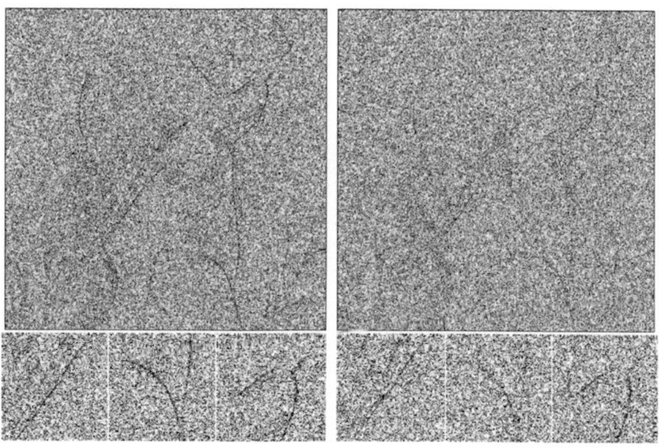

FIGURE 5.12: Close-ups of the clean curvature (left) and noisy curvature (right) from Figures 5.11(b) and 5.11(d) respectively, with non-linear scaling to highlight the differences.

Figure 5.14 shows, on the left, the noisy image I, in the middle the result u, the solution of (5.2), and on the right the original clean image a. The images u and a look very much alike, although there are slight numerical differences among them (the Mean Squared Error, MSE, between both images is 3.7).

In addition to the above observations, in [99] the authors proposed a variational approach for fusing a set of exposure bracketed images (a set of images of the same scene taken in rapid succession with different exposure times) that had a related, and initially somewhat perplexing, denoising effect. The energy functional fuses the colors of a long exposure image, I_{LE}, with the details from a short exposure image, I_{SE}, while attenuating noise from the latter. The denoising effect is surprisingly similar to that produced by state-of-the-art techniques directly applied to I_{SE}, such as Non-Local Means [109]. The term

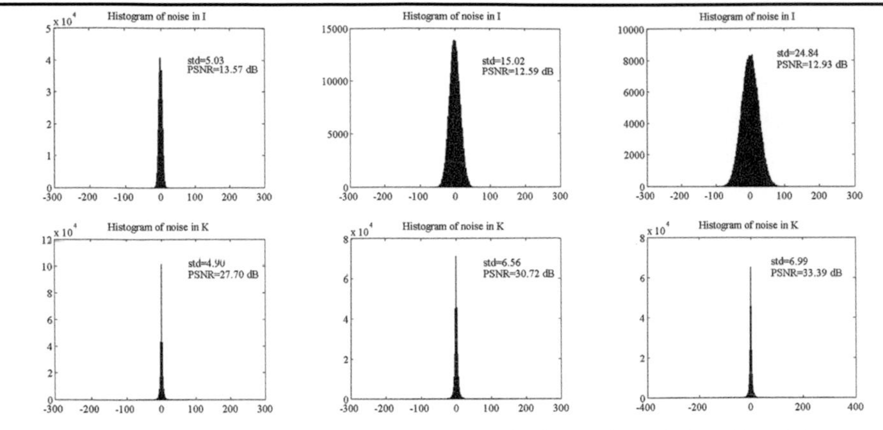

FIGURE 5.13: Noise histograms for I (top) and $\kappa(I)$ (bottom). From left to right: $\sigma = 5, 15, 25$.

FIGURE 5.14: Left: noisy image I. Middle: the result u obtained with (5.2). Right: original clean image a.

in the energy functional that generates this effect is $\int \left(|\nabla u| - \nabla u \cdot \frac{\nabla I_{SE}}{|\nabla I_{SE}|} \right)$ which was initially intended to preserve the details (i.e. gradient direction) of I_{SE}. The flow of the corresponding Euler-Lagrange equation for this term, $u_t = \kappa(u) - \kappa_\epsilon(I_{SE})$, is very similar to (5.2). Here $\kappa_\epsilon(I_{SE})$ is the curvature of I_{SE} which is computed using a small positive constant ϵ to avoid division by zero, and hence we could say that it has a regularizing effect on the actual curvature $\kappa(I_{SE})$; we found that as ϵ increases the final output of the fusion process becomes less noisy, therefore $\kappa_\epsilon(I_{SE})$ appears to be playing the role of the curvature of the clean image.

Motivated by the preceding observations, we propose the following general denoising framework. Given a noisy image $I = a + n$, instead of directly denoising I with some algorithm $\mathcal{F}$ to obtain a denoised image $I_{\mathcal{F}} = \mathcal{F}(I)$, do the following:

- Denoise the curvature image $\kappa(I)$ with method $\mathcal{F}$ to obtain $\kappa_{\mathcal{F}} = \mathcal{F}(\kappa(I))$.
- Generate an image $\hat{I}_{\mathcal{F}}$ that satisfies the following criteria:
  1. $\kappa(\hat{I}_{\mathcal{F}}) \simeq \kappa_{\mathcal{F}}$; that is, the level lines of $\hat{I}_{\mathcal{F}}$ are well described by $\kappa_{\mathcal{F}}$.
  2. The overall contrast of $\hat{I}_{\mathcal{F}}$ matches that of the given data $I = a+n$ in the sense that the intensity of any given level line of $\hat{I}_{\mathcal{F}}$ is close to the average value of I along that contour.

The resulting image $\hat{I}_{\mathcal{F}}$ described above will be a clean version of I, and one that we claim will generally have a higher PSNR and Q-index [372] than $I_{\mathcal{F}}$. It is important to point out that what we propose here is *not* necessarily a PDE-based denoising method, but rather a general denoising framework.

5.6.2 Comparing the noise power in I and in its curvature image $\kappa(I)$

5.6.2.1 PSNR along image contours

From (5.1) and basic calculus, the curvature of I can be written

$$\kappa(I) = \nabla \cdot \left(\frac{\nabla I}{|\nabla I|}\right) = \kappa(a)\frac{|\nabla a|}{|\nabla I|} + \frac{\nabla a}{|\nabla a|} \cdot \nabla \left(\frac{|\nabla a|}{|\nabla I|}\right) + \nabla \cdot \left(\frac{\nabla n}{|\nabla I|}\right). \tag{5.3}$$

First we consider the situation where

$$|\nabla a| \gg |\nabla n|, \tag{5.4}$$

which is likely the case at image contours. At an edge, where (5.4) holds, we have that $\frac{|\nabla a|}{|\nabla I|} \simeq 1$ and so the first term of the right-hand side of Equation (5.3) can be approximated by $\kappa(a)$, the second term $\frac{\nabla a}{|\nabla a|} \cdot \nabla \left(\frac{|\nabla a|}{|\nabla I|}\right) \simeq 0$ so it can be discarded (except in the case where the image contour separates perfectly flat regions, a scenario we discuss in Section 5.6.2.2), and finally the third term $\nabla \cdot \left(\frac{\nabla n}{|\nabla I|}\right)$ remains unchanged and is the main source of noise in the curvature image. So for now we approximate

$$\kappa(I) \simeq \kappa(a) + \nabla \cdot \left(\frac{\nabla n}{|\nabla I|}\right), \tag{5.5}$$

and consider the difference between the curvatures of the original and observed images in (5.5) as "curvature noise"

$$n_\kappa = \nabla \cdot \left(\frac{\nabla n}{|\nabla I|}\right). \tag{5.6}$$

In what follows, we approximate the curvature $\kappa(I)$ and unit normal field $\eta(I) = (\eta^1, \eta^2)$ of the image I using forward-backward differences, so

$$\kappa(I(x,y)) \simeq \Delta_-^x \left(\frac{\Delta_+^x I(x,y)}{|\nabla I(x,y)|}\right) + \Delta_-^y \left(\frac{\Delta_+^y I(x,y)}{|\nabla I(x,y)|}\right) \tag{5.7}$$

and

$$\vec{\eta}(I(x,y)) = (\eta^1(x,y), \eta^2(x,y)) \simeq \left(\frac{\Delta_+^x I(x,y)}{|\nabla I(x,y)|}, \frac{\Delta_+^y I(x,y)}{|\nabla I(x,y)|}\right), \tag{5.8}$$

where

$$\Delta_\pm^x I(x,y) = \pm\left(I(x \pm 1, y) - I(x,y)\right),\ \Delta_\pm^y I(x,y) = \pm\left(I(x, y \pm 1) - I(x,y)\right)$$

and where the discrete gradient is implied we use forward differences, so

$$|\nabla I(x,y)| = \sqrt{(\Delta_+^x I(x,y))^2 + (\Delta_+^y I(x,y))^2 + \epsilon^2}$$

for a small $\epsilon > 0$. In this setting, we have the following.

Proposition 1. *At locations in the image domain where $I = a + n$ satisfies (5.4) and (5.5) (likely the case at contours of I), and where the noise standard deviation satisfies $\sigma > \frac{|\nabla I|}{10.32}$, if the curvature $\kappa(I)$ is approximated by (5.7), then*

$$PSNR(I) < PSNR(\kappa).$$

Furthemore, if the unit normal field $\eta(I) = (\eta^1, \eta^2)$ is approximated by (5.8) and $\sigma > \frac{|\nabla I|}{3.64}$, then for $i = 1, 2$ we also have

$$PSNR(I) < PSNR(\eta^i) < PSNR(\kappa).$$

Proof. First we approximate the Peak Signal to Noise Ratio (PSNR) of $\kappa(I)$. Assuming I lies in the range $[0, 255]$ and that $\kappa(I)$ is computed using directional differences as described in (5.7), we have that $|\kappa| \leq 2 + \sqrt{2}$ and therefore the amplitude of the signal $\kappa(I)$ is $4 + 2\sqrt{2}$.

To compute $Var(n_\kappa)$, first observe that

$$n_\kappa = \nabla \cdot \left(\frac{n_x}{|\nabla I|}, \frac{n_y}{|\nabla I|}\right) = \left(\frac{n_x}{|\nabla I|}\right)_x + \left(\frac{n_y}{|\nabla I|}\right)_y. \tag{5.9}$$

Using forward-backward differences as described in (5.7) we have that

$$\begin{aligned}
\left(\frac{n_x}{|\nabla I|}\right)_x &\simeq \Delta_-^x \left(\frac{\Delta_+^x (n(x,y))}{|\nabla I(x,y)|}\right) \qquad (5.10)\\
&= \frac{\Delta_+^x (n(x,y))}{|\nabla I(x,y)|} - \frac{\Delta_+^x (n(x-1,y))}{|\nabla I(x-1,y)|}\\
&= \frac{\Delta_+^x (n(x,y))|\nabla I(x-1,y)| - \Delta_+^x (n(x-1,y))|\nabla I(x,y)|}{|\nabla I(x,y)||\nabla I(x-1,y)|}\\
&= \frac{\Delta_-^x (\Delta_+^x (n(x,y)))}{|\nabla I(x,y)|} - \frac{\Delta_+^x (n(x-1,y))\Delta_-^x |\nabla I(x,y)|}{|\nabla I(x,y)||\nabla I(x-1,y)|}
\end{aligned}$$

Without loss of generality, assume the edge is vertical, so $I_y \simeq 0$. If the edge discontinuity occurs between x and $x+1$, then

$$|\nabla I(x-1,y)| \simeq |\Delta_+^x I(x-1,y)| \simeq |\Delta_+^x n(x-1,y)|$$

and thus

$$\Delta_-^x |\nabla I(x,y)| = |\nabla I(x,y)| - |\nabla I(x-1,y)| \simeq |\nabla I(x,y)| - |\Delta_+^x n(x-1,y)|.$$

From the above calculations, the second term on the right hand side of (5.10) satisfies

$$\frac{\Delta_+^x(n(x-1,y))\Delta_-^x|\nabla I(x,y)|}{|\nabla I(x,y)||\nabla I(x-1,y)|} \simeq \pm \frac{|\nabla I(x,y)| - |\Delta_+^x n(x-1,y)|}{|\nabla I(x,y)|} \tag{5.11}$$

which is bounded above by 1 due to (5.4). Since an upper bound is sufficient for our argument, by (5.10) an (5.11) we can approximate

$$\left(\frac{n_x}{|\nabla I|}\right)_x \simeq \frac{\Delta_-^x(\Delta_+^x(n(x,y)))}{|\nabla I(x,y)|} + T_x, \quad \text{where} \quad T_x \in [0,1]. \tag{5.12}$$

Similar to (5.10),

$$\left(\frac{n_y}{|\nabla I|}\right)_y \simeq \frac{\Delta_-^y(\Delta_+^y(n(x,y)))}{|\nabla I(x,y)|} - \frac{\Delta_+^y(n(x,y-1))\Delta_-^y|\nabla I(x,y)|}{|\nabla I(x,y)||\nabla I(x,y-1)|}.$$

At a vertical edge we would expect that $|\nabla I(x,y)| \simeq |\nabla I(x,y-1)| >> \Delta_+^y(n(x,y-1))$ and $\Delta_-^y|\nabla I(x,y)| \simeq 0$. Therefore

$$\left(\frac{n_y}{|\nabla I|}\right)_y \simeq \frac{\Delta_-^y(\Delta_+^y(n(x,y)))}{|\nabla I(x,y)|}. \tag{5.13}$$

By (5.9), (5.12), and (5.13) we have that

$$\begin{aligned} n_\kappa &\simeq \frac{\Delta_-^x(\Delta_+^x(n(x,y)))}{|\nabla I(x,y)|} + \frac{\Delta_-^y(\Delta_+^y(n(x,y)))}{|\nabla I(x,y)|} + T_x \qquad (5.14)\\ &= \frac{1}{|\nabla I|}(n(x+1,y) + n(x-1,y) + n(x,y+1) + n(x,y-1) - 4n(x,y)) + T_x. \end{aligned}$$

Assuming $n \sim \mathcal{N}(0, \sigma^2)$, the (numerical) variance of n_κ is then

$$\begin{aligned}
&Var(n_\kappa) \\
&\simeq \frac{Var(n(x+1,y) + n(x-1,y) + n(x,y+1) + n(x,y-1))}{|\nabla I|^2} + \\
&+ \frac{16Var(n(x,y))}{|\nabla I|^2} + Var(T_x) \\
&= \frac{1}{|\nabla I|^2}(4Var(n) + 16Var(n)) + Var(T_x) \\
&= \frac{1}{|\nabla I|^2}20Var(n) + Var(T_x) \\
&= \frac{20}{|\nabla I|^2}\sigma^2 + Var(T_x).
\end{aligned}$$

Therefore, we typically have that

$$Var(n_\kappa) \simeq \frac{20}{|\nabla I|^2}\sigma^2 + Var(T_x) \quad \text{where} \quad Var(T_x) \in [0, 0.25]. \tag{5.15}$$

Now we can compute the PSNR of $\kappa(I)$, as the peak amplitude of the curvature signal is $4 + 2\sqrt{2}$ and the variance of the noise is given by (5.15), so

$$PSNR(\kappa(I)) \simeq 20log_{10}\left(\frac{4 + 2\sqrt{2}}{\sqrt{\frac{20\sigma^2}{|\nabla I|^2} + Var(T_x)}}\right). \tag{5.16}$$

Since $Var(T_x) \in [0, 0.25]$, at locations where $\sigma > \frac{|\nabla I|}{10.32}$ we have that

$$PSNR(\kappa(I)) \in \left(20log_{10}\left(\frac{|\nabla I|}{\sigma}\right), 20log_{10}\left(1.53\frac{|\nabla I|}{\sigma}\right)\right] \tag{5.17}$$

If we go to the original grayscale image I and compute *locally* its PSNR, we get that the amplitude is approximately $|\nabla I|$ (because the local amplitude is the magnitude of the jump at the boundary, and using directional differences $|\nabla I|$ is the value of this jump) and the standard deviation of the noise is just σ, therefore

$$PSNR(I) = 20log_{10}\left(\frac{|\nabla I|}{\sigma}\right). \tag{5.18}$$

This would be saying that, along the contours of a, the curvature image $\kappa(I)$ will be up to 3.7dB *less noisy* than the image I.

What happens if we want to denoise the normals, as in Lysaker et al. [260]? Let $\vec{\eta}$ be the normal vector

$$\vec{\eta} = (\eta^1, \eta^2) = \frac{\nabla I}{|\nabla I|} = \frac{\nabla a}{|\nabla I|} + \frac{\nabla n}{|\nabla I|}. \tag{5.19}$$

Let's compute the PSNR for any of the components of $\vec{\eta}$, say η^1. Its amplitude is 2, since $\eta^1 \in [-1, 1]$. Using similar arguments as before, we can approximate the variance of the "noise" in η^1 as

$$Var(\frac{n_x}{|\nabla I|}) \simeq \frac{1}{|\nabla I|^2} Var(n_x), \tag{5.20}$$

and, using directional differences

$$n_x(x,y) = n(x+1,y) - n(x,y), \tag{5.21}$$

so

$$Var\left(\frac{n_x}{|\nabla I|}\right) \simeq \frac{1}{|\nabla I|^2} 2Var(n) = \frac{1}{|\nabla I|^2} 2\sigma^2. \tag{5.22}$$

Therefore, the PSNR of the first component of the normal field is

$$PSNR(\eta^1) = 20log_{10}\left(\frac{2}{\sqrt{2}\frac{\sigma}{|\nabla I|}}\right) = 20log_{10}\left(1.41\frac{|\nabla I|}{\sigma}\right). \tag{5.23}$$

If $\sigma > \frac{|\nabla I|}{3.64}$ then

$$PSNR(\kappa(I)) \in \left(20log_{10}\left(\frac{1.41|\nabla I|}{\sigma}\right), 20log_{10}\left(1.53\frac{|\nabla I|}{\sigma}\right)\right] \tag{5.24}$$

From (5.18), (5.23) and (5.24) we get $PSNR(I) < PSNR(\eta^i) < PSNR(\kappa)$. □

Remark 1. *The restrictions on σ in Proposition 1 are fairly conservative given the experimental results that follow in Section 5.6.2.3 and Section 5.6.4 (e.g. see Figures 5.15 and 5.16). But the overall conclusion is still the same, so we included these hypotheses for ease of argument.*

Remark 2. *Note that if instead of using forward-backward differences to compute κ we had used central differences and the formula*

$$\kappa = \frac{I_x^2 I_{yy} + I_y^2 I_{xx} - 2I_x I_y I_{xy}}{(I_x^2 + I_y^2)^{\frac{3}{2}}},$$

then the amplitude of κ would be much larger than $4 + 2\sqrt{2}$ and hence the difference in PSNR with respect to I would also be much larger. But we have preferred to consider the case of directional differences, because in practice the curvature is usually computed this way, for numerical stability reasons (see Ciomaga et al. [129] for alternate ways of estimating the curvature).

The above conclusions suggest that, given any denoising method, for best results *on the contours* it may be better to denoise the curvature rather than directly denoise I (or the normal field).

5.6.2.2 Correction for contours separating flat regions

As we mentioned earlier, if we have an image contour that separates perfectly flat regions then the second term of the right-hand side of Equation (5.3) cannot be discarded. The reason is that while $\frac{|\nabla a|}{|\nabla I|} \simeq 1$ holds on the contour, we also have $\frac{|\nabla a|}{|\nabla I|} \simeq 0$ on its sides because these regions are flat (and hence $|\nabla a| \simeq 0$). Consequently, we can no longer approximate the term $\frac{\nabla a}{|\nabla a|} \cdot \nabla \left(\frac{|\nabla a|}{|\nabla I|}\right)$ by zero, but we can bound its variance.

Consider a 100×100 square image a with value 0 for all pixels in the columns $0 - 49$ and value 255 for all pixels in the columns $50 - 99$. Image a then has a vertical edge that separates flat regions. Using backward differences, the term $\frac{|\nabla a|}{|\nabla I|} \simeq 0$ everywhere except at column 50, where $\frac{|\nabla a|}{|\nabla I|} \simeq 1$. Therefore, the term $\nabla \left(\frac{|\nabla a|}{|\nabla I|}\right)$ is close to $(0, 0)$ everywhere except at column 50, where it is $(1, 0)$, and column 51, where it is $(-1, 0)$ (always approximately). So the second term of the right-hand side of Equation (5.3), $\frac{\nabla a}{|\nabla a|} \cdot \nabla \left(\frac{|\nabla a|}{|\nabla I|}\right)$, is close to zero everywhere except at column 50, where it is close to 1. Exactly the same result holds if the image a is flipped and takes the value 255 on the left and 0 on the right, because now $\nabla \left(\frac{|\nabla a|}{|\nabla I|}\right)$ is approximately $(-1, 0)$ at column 50 but there the normalized gradient $\frac{\nabla a}{|\nabla a|} \simeq (-1, 0)$ as well.

The conclusion is that, in practice, the second term of the right-hand side of Equation (5.3) is in the range $[0, 1]$, so we may bound its variance by 0.25. This leads to a correction of Equation (5.15) for this type of contour

$$Var(n_\kappa) \simeq \frac{20}{|\nabla I|^2}\sigma^2 + T'_x \quad \text{where} \quad T'_x \in [0, 0.5]. \tag{5.25}$$

5.6.2.3 PSNR along contours: numerical experiments

We have performed tests on two very simple synthetic images, one binary and the other textured, where we add noise with different σ values to them and compute the PSNR of the image, curvature and normal field along the central circumference (for the normal we average PSNR values of the vertical and horizontal components).

Figure 5.15 shows the results for the textured image, where we can see that the PSNR values are consistent with our estimates.

Figure 5.16 shows the results for the binary image, which are also consistent with our estimates once we introduce the correction term of Equation (5.25). As the equation predicts, for this case we see that if σ is small then the PSNR along the contours of the image may be larger than that of the curvature. Nonetheless, this does not affect the results of our denoising framework, which we will detail in Section 5.6.3: with our approach we obtain denoised results with higher PSNR, computed over the whole image, even for binary images and small values of σ. In particular, for the binary circle image of Figure 5.16, for noise of standard deviation $\sigma = 5$ and for total variation (TV) based denoising with $\mathcal{F} = ROF$ [330] we obtain, with our proposed framework (i.e. by applying TV denoising to the curvature), a denoised image result with PSNR=47.85, whereas direct TV denoising on the image yields PSNR=46.77. The influence of homogeneous regions on the PSNR is discussed next.

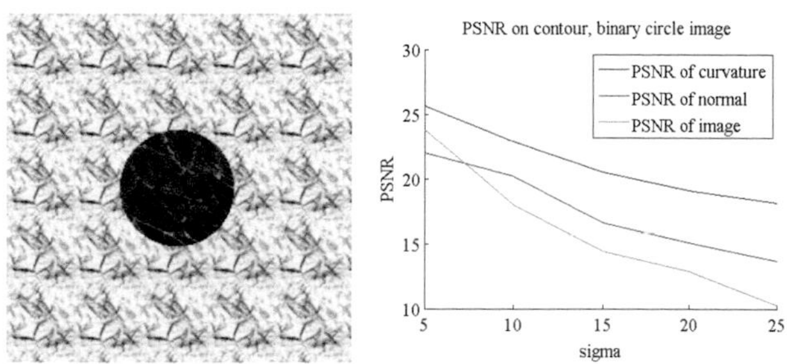

FIGURE 5.15: Left: test image. Right: PSNR values of image, curvature and normal along contour.

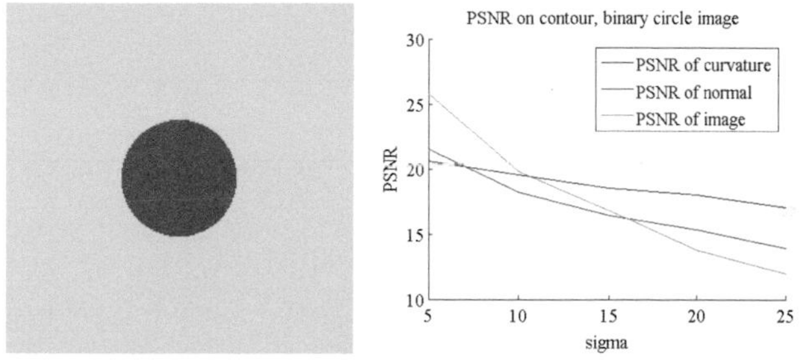

FIGURE 5.16: Left: test image. Right: PSNR values of image, curvature and normal along contour.

5.6.2.4 PSNR in homogeneous regions

On homogeneous or slowly varying regions, (5.4) is no longer valid and we have instead

$$|\nabla a| \ll |\nabla n|, \tag{5.26}$$

so now

$$\kappa(I) \simeq \kappa(n) + \nabla \cdot (\frac{\nabla a}{|\nabla I|}). \tag{5.27}$$

In this case $\kappa(I)$ cannot be expressed as the original curvature $\kappa(a)$ plus some *curvature noise*, unlike in (5.5). So in homogeneous regions $\kappa(I)$ is a poor estimation of $\kappa(a)$, but we can argue that this is not a crucial issue, with the following reasoning.

From (5.26) and (5.27) we see that $\kappa(I)$ behaves like $\kappa(n)$ plus a perturbation. Since n is random noise with mean zero, so is $\kappa(n)$ and thus so is $\kappa(I)$. Therefore, any simple denoising method applied to $\kappa(I)$ will result in values of $\kappa_{\mathcal{F}}$ close to zero in homogeneous or slowly varying regions. So after running *Step 2* of the proposed approach below in Algorithm 2 (for e.g. we could use (5.2)), the reconstructed (denoised) image $\hat{I}_{\mathcal{F}}$ will have, in these homogeneous regions, curvature close to zero, which means that these regions will be approximated by planes (not necessarily horizontal). This is not a bad approximation given that these regions are, precisely, homogeneous or slowly varying.

5.6.3 Proposed algorithm

5.6.3.1 The model

The observations in the previous sections have motivated us to perform a number of experiments comparing the following two algorithms.

Algorithm 1 Direct approach

Apply a denoising approach $\mathcal{F}$ to directly smooth an image I, obtaining a denoised image $I_{\mathcal{F}} = \mathcal{F}(I)$.

Algorithm 2 Proposed approach

Step 1: Given a noisy image, I, denoise $\kappa(I)$ with method $\mathcal{F}$ to obtain $\kappa_{\mathcal{F}} = \mathcal{F}(\kappa(I))$.

Step 2: Generate an image $\hat{I}_{\mathcal{F}}$ that satisfies the following criteria:

1. $\kappa(\hat{I}_{\mathcal{F}}) \simeq \kappa_{\mathcal{F}}$; that is, the level lines of $\hat{I}_{\mathcal{F}}$ are well described by $\kappa_{\mathcal{F}}$.
2. The overall contrast of $\hat{I}_{\mathcal{F}}$ matches that of the given data $I = a + n$ in the sense that the intensity of any given level line of $\hat{I}_{\mathcal{F}}$ is close to the average value of I along that contour.

We have tested both variational and patch based approaches for the denoising method $\mathcal{F}$. So $\kappa_{\mathcal{F}}$ has been generated from fairly diverse methods in *Step 1*.

The precise method of reconstruction for *Step 2* should potentially be related to the nature of the smoothed curvature $\kappa_{\mathcal{F}}$ from *Step 1*, and thus the choice of denoising method $\mathcal{F}$ as well as the discretization of $\kappa(I)$. For simplicity, for all of the tests in this paper we have performed *Step 2* by solving

$$u_t = \kappa(u) - \kappa_{\mathcal{F}} + 2\lambda(I - u), \tag{5.28}$$

with initial data $u(0, \cdot) = I$ or $u(0, \cdot) = I_{\mathcal{F}}$ where λ is a positive parameter depending on the noise level (and possibly depending on time). This is just one choice and in [100] we discuss other alternatives. But we chose to use (5.28) as a baseline for our experiments since its behavior is well-understood.

5.6.4 Experiments

The image database used in our experiments is the set of grayscale images (range $[0, 255]$) obtained by computing the luminance channel of the images in the Kodak [231] database (at half-resolution). We tested four denoising methods: TV denoising [330], the Bregman iterative algorithm [297], orientation matching using smoothed unit tangents [191], and Non-local Means [109]. Our experiments show that for all of these algorithms, we obtain better results by denoising the curvature image $\kappa(I)$ rather than directly denoising the image I.

To compute $\kappa(u)$ in the reconstruction Equation (5.28) we have used the classical numerical scheme of [330], with forward-backward differences and the minmod operator, to ensure stability. Therefore, we also use this for the initialization of the noisy curvature $\kappa(I)$.

FIGURE 5.17: Left: noisy image. Middle: result obtained with TV denoising of the image (PSNR=29.20 and PIQ=82.45). Right: result obtained with TV denoising of the curvature image (PSNR=29.36 and PIQ=95.64). From [100].

Figure 5.17 shows one example comparing the outputs of TV denoising of I and $\kappa(I)$ for the Lena image and noise with $\sigma = 25$. It is useful to employ for image quality assessment, apart from the PSNR, the Q-index of [372], which is reported as having higher perceptual correlation than PSNR and SNR-based metrics [306]; in our case we use the percentage increase in Q,

$$PIQ(I_{ROF}) = 100 \times \frac{Q(I_{ROF}) - Q(I)}{Q(I)},$$

$$PIQ(\hat{I}_{ROF}) = 100 \times \frac{Q(\hat{I}_{ROF}) - Q(I)}{Q(I)}.$$

In this image we obtain PSNR=29.36 and PIQ=95.64 for TV denoising of the curvature, while the values are PSNR=29.20 and PIQ=82.45 for TV denoising of the image.

Figure 5.18 compares, on the left, the average increase in PSNR, computed over the entire Kodak database, obtained with both approaches: PSNR(I_{ROF})-PSNR(I) (in magenta), PSNR($\hat{I}_{ROF}$)-PSNR(I) (in blue). On the right, we plot the average percentage increase in Q-index. Both plots in Figure 5.18 show that TV denoising of the curvature allows us to obtain a denoised image $\hat{I}_{ROF}$ which is better in terms of PSNR and Q-index than I_{ROF}, the image obtained by directly applying TV denoising to the original noisy image.

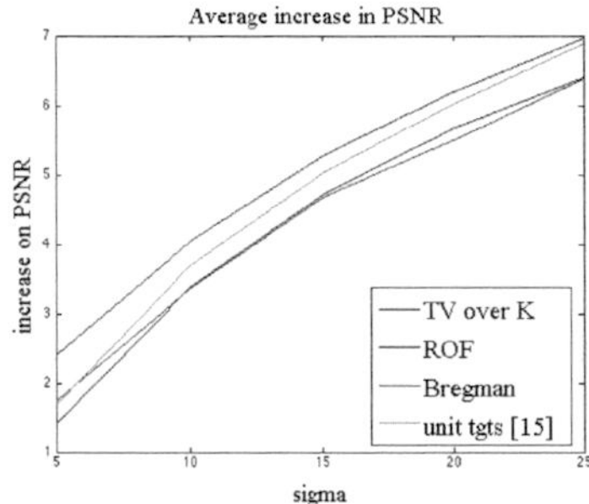

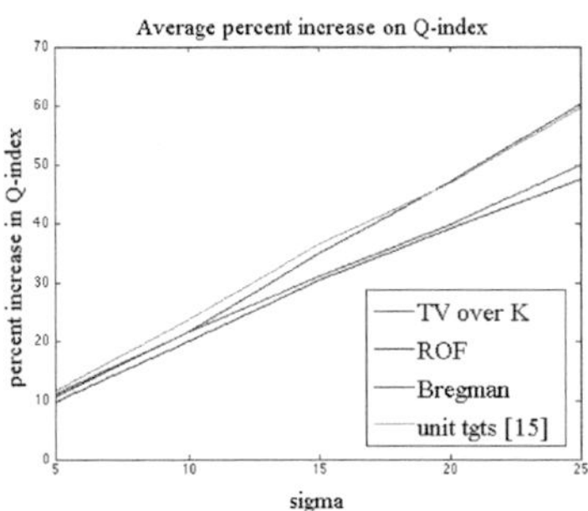

FIGURE 5.18: Comparison of TV denoising of $\kappa(I)$ (blue), smoothing unit tangents [191] (green), TV denoising of I (ROF) (magenta), the Bregman iterative approach (red). Left: PSNR increase for each method. Right: percentage increase on Q-index [372]. Values averaged over Kodak database (only luminance channel, images reduced to half-resolution). From [100].

FIGURE 5.19: Left: noisy image. Middle: result obtained with NLM denoising of the image. Right: result obtained with NLM denoising of the curvature image. From [100].

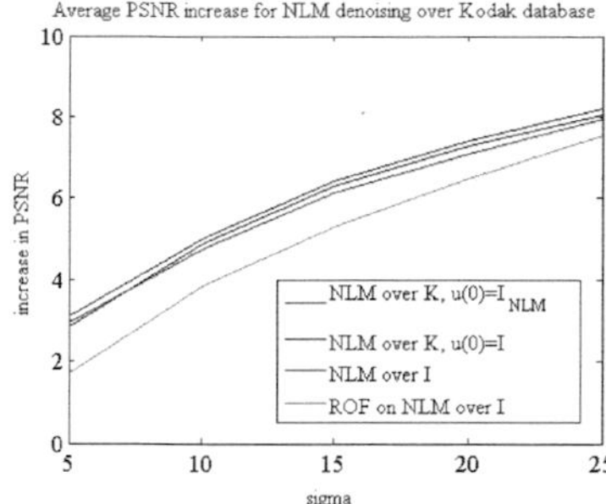

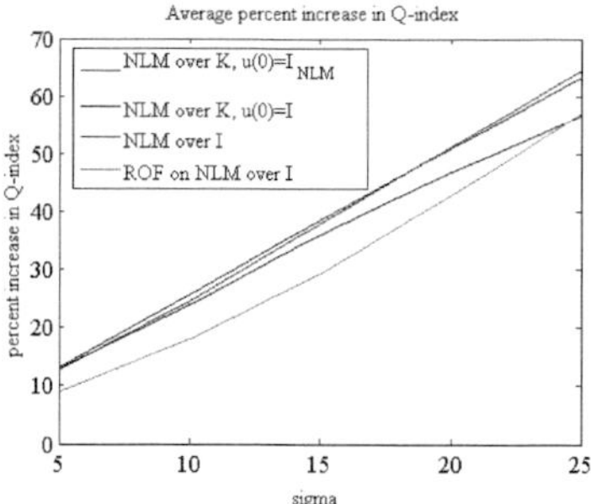

FIGURE 5.20: Comparison of NLM denoising on I and NLM denoising on $\kappa(I)$. Left: PSNR increase for each method; also pictured: PSNR increase for ROF applied to the output of NLM on I. Right: percentage increase on Q-index [372]. Values averaged over Kodak database (only luminance channel, images reduced to half-resolution). From [100].

Figure 5.19 shows one example comparing the outputs of NLM denoising of I and $\kappa(I)$. Figure 5.20 (left) compares the average increase in PSNR over the entire Kodak database of the denoised image over the original noisy image, obtained with both approaches: NLM applied to I (in magenta) and NLM applied to κ with different initial conditions (blue and green).

Both plots in Figure 5.20 show that NLM denoising of the curvature allows us to obtain a denoised image $\hat{I}_{NLM}$ which is better in terms of PSNR and Q-index than the image I_{NLM}, obtained directly by applying NLM to the original noisy image.

Chapter 6

Demosaicking and deinterlacing

6.1 Introduction

We have seen in Chapter 3 that image sensors measure light intensity but not its wavelength, and that in order to capture colors the most common configuration is to use a single sensor covered with a mosaic color filter called Color Filter Array (CFA). By far the most popular CFA is the Bayer pattern, using an RGB color choice with two green pixels per red or blue pixel; see Figure 6.1.

FIGURE 6.1: Bayer CFA.

Therefore each pixel only has information of a single color channel, and the other two values must be interpolated through a process called demosaicking or demosaicing. Basic interpolation techniques do not yield good results; see Figure 6.2, and more sophisticated methods may also fail, producing artifacts such as false colors or the so-called zipper-effect, an on-off pattern appearing near strong edges and areas with saturated colors; see Figure 6.5. Therefore this is a challenging problem and there is extensive literature about it; the interested reader is referred to the recent survey by Menon and Calvagno [273]. In this chapter we will give a brief overview of the most relevant types of techniques that can be used for demosaicking.

While demosaicking is an interpolation procedure that must always be performed right at the beginning, either on camera or off-line, when images are captured with a single-sensor camera, there is in some cases the need to carry out another type of interpolation also early on in the image processing chain

FIGURE 6.2: Problems with demosaicking. Left: full-color image used as ground truth. Only one color channel per pixel is kept, and this simulated mosaicked picture is demosaicked with linear interpolation, yielding the result on the right.

of the post-production, and that is deinterlacing. Current day professional footage is never shot in interlaced mode, but amateur video or legacy material might be in this form in which case an interlaced to progressive conversion is necessary. As always, simple methods don't work: fusing two fields in a single frame (Figure 6.3 left) produces noticeable artifacts on moving regions, and averaging consecutive lines in the same field (Figure 6.3 right) blurs static elements. The second section of this chapter will present some deinterlacing algorithms that are relevant for cinema postproduction, since they provide high quality results without compromises for real time implementation (which are required for real-time deinterlacing in TV sets, for instance).

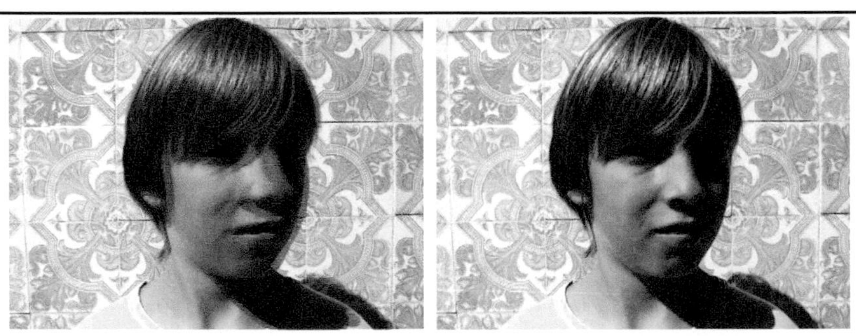

FIGURE 6.3: Problems with deinterlacing. Left: field weaving result. Right: line average result.

6.2 Demosaicking

In a representative frequency domain approach, Lian et al. [251] start by making two observations: first, that high frequencies are similar across color channels; second, that it's better to preserve high frequencies along the horizontal and vertical axes, where the subsampled green channel is not so much aliased as the other two components. They design a low-pass filter that is applied to the luminance but with results restricted to the green channel. For the full-resolution luminance an adaptive-filtering approach is used, computing for each pixel a weighted average of color differences in a neighborhood, where the weights are adapted to the orientation of local edges.

Zhang and Wu [400] note that in the demosaicking literature two assumptions are made regarding the color differences green-red and green-blue: equal ratio (for data prior to gamma correction) and constant ratio (for data after gamma correction). In [400] the authors assume that the difference images are low-pass, which leads to an adaptive estimation of the missing green values in the horizontal and vertical directions using a relaxation of the minimum mean square error (MMSE) technique. These estimates are combined through a weighted average, where the weights are chosen to minimize the MSE of the result. Finally, the full red and blue channels are generated from the full green channel and the difference images.

The approach of the previous work, where both a vertical and a horizontal interpolation are produced and then combined, is quite common, and other works follow it with different criteria for the combination procedure. Hirakawa and Parks [196] propose to follow the direction with less image artifacts. Chung and Chan [128] look for the direction (horizontal, vertical or diagonal) of minimum variance of the color differences, and interpolate the green channel along this direction. Tsai and Song [361] and Menon et al. [271] suggest interpolating the missing color information along the direction of image contours, and hence they are edge-direction approaches, albeit local because the estimation of the direction of the contour is performed in a local neighborhood. Ferradans et al. [163] propose a non-local edge-direction method, posed as an inpainting problem: each image diagonal without green samples is considered as an image gap that must be inpainted, using the information from the adjacent diagonals. The inpainting problem is solved by minimizing a functional that measures the cost of matching the neighboring diagonals, and the matches give the resulting interpolation directions, which change locally but are estimated through a global minimization procedure. See Figure 6.4. Figure 6.5 compares the results of several of the abovementioned algorithms.

Possibly the first instance of a patch-based, non-local demosaicking algorithm was the work of Mairal et al. [261], that introduced a general framework of sparse representation for several image restoration problems including, aside

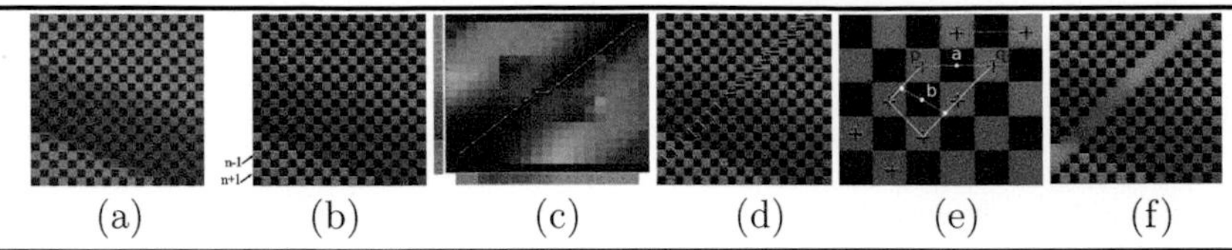

FIGURE 6.4: Schematic for the method of Ferradans et al. [163]. (a) Image mosaicked with the Bayer filter. (b) Green plane, note the 45° diagonals $n-1$, $n+1$. (c) Cost matrix for diagonals $n-1$ and $n+1$. (d) Dense set of correspondences (points in the optimal path) obtained between diagonals $n-1$ and $n+1$ with optimal path (minimum accumulated cost) superimposed in red. (e) Interpolation process; note the different cases for pixel a and pixel b. (f) Complete diagonal obtained by interpolation. Figure from [163].

from demosaicking, image denoising and image inpainting. This approach is an extension of the K-SVD denoising algorithm mentioned in Chapter 5:

- An image dictionary of patches is built, and each patch of the noisy image is expressed as a linear combination from the dictionary patches, plus an error.
- A triply-nested optimization procedure looks for the smallest dictionary where the noisy image patches are a combination of the smallest number of dictionary patches and produce the smallest errors.
- The output of the optimization procedure is the denoised image.

For demosaicking, the algorithm is modified and the problem slightly reformulated as an inpainting problem with very small holes consisting of two channels per pixel. In order to prevent the (modified) K-SVD algorithm from learning the hole pattern (i.e. to prevent that this pattern ends up appearing in the dictionary patches) the authors employ an adaptive dictionary that has been learned with low patch sparsity, although they also mention that a globally learned dictionary with no-mosaic images could be used.

Buades et al. [111] adapt their non-local means (NLM) denoising formulation [109] to the demosaicking problem. We recall that in the NLM algorithm the value at each pixel is replaced by a weighted average of all pixels in the image, weighting them according to the similarity of their patch neighborhoods. Now, in the demosaicking context, they start by using a classical interpolation scheme to produce a first estimate, which might be blurry and suffering from several types of artifacts. Then they apply to this image a variant of the NLM algorithm where the pixels averaged are only those whose values are known, and these averages are transported only to unknown values. The algorithm works in a coarse-to-fine fashion by progressively decreasing the value of the resolution parameter h (which in the case of denoising was selected according

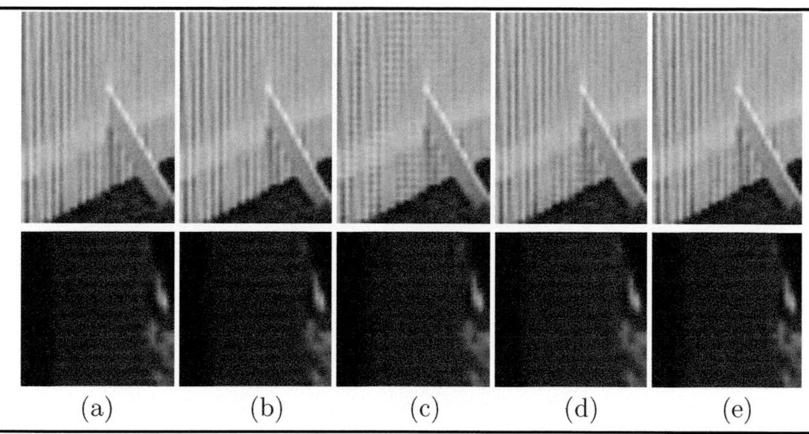

FIGURE 6.5: Demosaicking results for several methods. (a) Ground truth. (b) Ferradans et al. [163]. (c) Lian et al. [251]. (d) Zhang et al. [400]. (e) Chung and Chen [128]. Figure from [163].

to the power of the noise): for a large value of h the results are blurry but reliable, and as h decreases the algorithm tends to transport values instead of performing averages; see Figure 6.6. This transport operation may cause impulse noise in case of error, so a post-processing step is added where a very local median filter is applied to the chromaticity components U, V of each pixel. Being non-local, the computational cost of this algorithm is comparable to that of other sophisticated methods but lower than that of [261], where a dictionary must be learnt for each image.

Gao et al. [179] recently proposed a method based on sparse representation. They apply an L_1 minimization approach to a combination of two sparse representations of mosaiced images: one characterizing spectral correlations (where color difference images are modeled as piecewise linear functions and therefore their Laplacians are sparse), the other characterizing spatial correlations and based on principal component analysis (PCA). For each pixel a patch centered in it is considered, this patch is matched to other patches in the image, and all the matching patches are stacked together so as to compute local PCA bases, in which the image signal is sparse. The authors prefer PCA over signal-independent bases like DCT or wavalets because PCA provides a better fit to local image structures. The numerical results reported are quite impressive, surpassing state-of-the-art algorithms in several dB's.

The use of PCA in the context of demosaicking had been proposed by Zhang et al. [399], for *denoising* of CFA images previous to the application of any demosaicking algorithm. PCA is applied to sets of similar image blocks, containing color values from different channels, and noise is reduced by removing PCA components with low energy. Actually the problems of denoising and

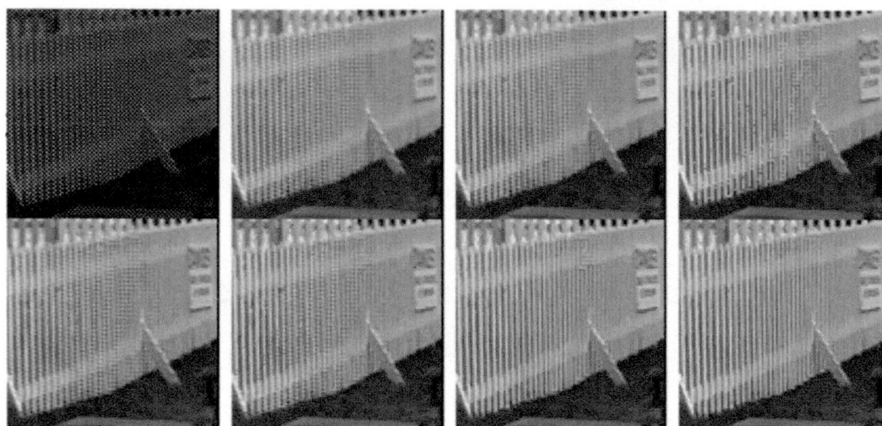

FIGURE 6.6: The demosaicking method of Buades et al. [111]. Gray interpolation experiment for various decreasing values of the parameter h. Top: initial image, bi-linear interpolation, median-based interpolation, anisotropic interpolation. Bottom: result of [111] for decreasing sequences of values for h. The initial condition is always given by the median. Figure from [111].

demosaicking are intimately related, and treating them simultaneously is advocated by several works, which argue that [273]:

- Denoising before demosaicking must be applied on the three color channels independently, thus not working at the full spatial resolution of the image and also ignoring the correlation among color channels, yielding a poorer denoising result.
- Demosaicking before denoising implies that the noise will be present and affecting the estimation of edge direction required by many demosaicking algorithms; also, the demosaicking procedure alters the statistics of the noise and this might affect the behavior of some denoising methods.

Hirakawa and Parks [197] were the first to jointly perform demosaicking and denoising. In their method each pixel value is replaced by a weighted average of its neighbors, and this is the main difference with demosaicking algorithms, which only fill in missing values but respect those present in the CFA image. The weights are computed by locally estimating an optimal filter, and there is room for different denosing techniques as long as an optimality constraint is satisfied.

Paliy et al. [299] compute color difference images from initial interpolated estimates, decorrelate them, and denoise them using directional 1D filters. At each pixel and for each direction the optimum scale (window size of the filter) is computed, and the final estimate is a linear combination of the directional estimates at optimum scales. From these values, all R, G and B components at every position can be computed. See Figure 6.7 for a comparison with the previous method.

Menon and Calvagno [272] perform an initial estimate of the full color image, to which they apply a wavelet transform followed by hard-thresholding of the coefficients (an operation that reduces the noise, as seen in Chapter 5), inverse-transform and finally an adaptive reconstruction of the color channels where the edge direction estimation is computed on the denoised channels.

FIGURE 6.7: Simultaneous denoising and demosaicking. Left: result of applying just demosaicking to a noisy CFA image. Middle: joint denoising/demosaicking result of [197]. Right: joint denoising/demosaicking result of [299]. Figure from [299].

Finally, let's mention some methods for video demosaicking. Wu and Zhang's algorithm [385] works in the following steps:

- The green channels of all frames are demosaicked individually.
- Motion estimation is performed on these interpolated green channel frames.
- Motion vectors are used to register frames and enhance green channel images by fusing corresponding values from different frames.
- In each image, the improved green channel is used to reconstruct the red and blue channels.
- The interpolated red and blue channels are refined by fusion with temporal matches from the other frames.

Zhang et al. [398] propose a method that starts by performing PCA-based denoising as in [399], but now the similar patches are sought both spatially and temporally. The denoised sequence is demosaicked on a frame-by-frame basis, and finally NLM is applied so as to remove color artifacts (again, patch comparison in NLM is now performed both intra- and inter-frame). The results of both [385] and [398] are a bit underwhelming, showing loss of detail, particularly in [398].

6.3 Deinterlacing

Deinterlacing algorithms can be broadly classified into four categories [84] which are, in increasing order of complexity:

1. Linear methods. Lines are interpolated by very simple operations such as line replication (called line doubling if done intra-frame, and field weaving if done temporally), or averaging (line average if done spatially, field averaging if done with temporally adjacent fields).

2. Directional interpolation methods. The interpolation procedure is carried out in the (estimated) direction of the spatio-temporal details.

3. Motion adaptive methods. Motion detection (not *estimation*) is used to switch between temporal interpolation (more accurate when there isn't motion) and spatial interpolation (in the presence of motion).

4. Motion compensated methods. Motion estimation is performed and the interpolation is conducted along the directions determined by the motion vector field.

The latter class of algorithms usually produces the best quality of results, at the expense of a much higher computational cost than simpler techniques such as directional interpolators. Current display technology produces progressive output, so for interlaced inputs the device must perform, in real time, the interlaced to progressive conversion. For that type of application, the demands on the algorithm performance are very high and the final choice of method is usually a compromise between image quality and algorithm complexity; see for instance [247] and references therein for a recent work of this type.

For cinema postproduction, though, visual quality is the paramount concern so the methods that are preferred are those that yield better results, pretty much ignoring the computational cost since the procedure will be carried out off-line and only once. We will mention just a few of these works providing high-quality deinterlacing.

Ballester et al. [84] pose the interlaced to progressive conversion as an inpainting problem, where each missing line is an "image gap" that must be filled in from its boundary, which are the four spatio-temporal neighboring lines (two in the same field (up and down), two at the same position in the previous and next fields). The inpainting is performed first intra- and then inter-frame, and both solutions are combined with a weighted average. The inpainting problem is solved by globally minimizing a functional that measures the cost of matching the two neighboring lines; the matches provide the direction for interpolation, and the process guarantees that lines (edges) do not cross. So this can be seen as a directional interpolation method, although the direction is estimated in a global way, not local as it's usually done. The algorithm is reported to outperform state-of-the-art motion compensated methods in terms of visual quality, and at a comparable computational cost. See Figure 6.8 for some example results.

FIGURE 6.8: Comparison of several deinterlacing methods. From left to right: original, results of [391], [295], [140] and [84]. Figure from [84].

Keller et al. [225] define several energy functionals that are minimized in order to produce a deinterlaced output. The stages of their method are the following:

1. Split video in even and odd field sets.
2. Perform optical flow estimation separately on each set by minimizing an energy with terms involving the regularity of the image, of its gradient and that of the motion vector field.

3. Perform fusion, scaling and smoothing of both optical flow fields into a single regularized field.

4. Use this motion field to perform motion compensated deinterlacing by minimizing another energy functional that takes into account the regularity of the solution and of its change along the estimated motion direction.

The results show a noticeable over-smoothing, due to the denoising behavior of the final energy minimization stage.

Ghodstinat et al. [181] propose a method that works in three steps:

1. Spatial deinterlacing based on an anisotropic, edge enhancing diffusion method previously used for interpolation.

2. Motion estimation with a state-of-the-art optical flow computation approach.

3. Image blocks are aligned with the motion field of *step 2* and 3D-filtered with the approach of *step 1.*

Figure 6.9 compares this method with the previous one [225].

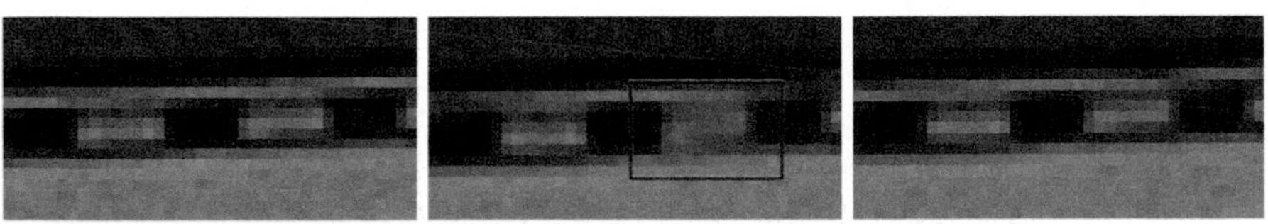

FIGURE 6.9: Comparison of deinterlacing algorithms. Left: original (ground truth). Middle: motion adaptive result of [225]. Right: result of [181]. Figure from [181].

Li and Zheng [250] propose a method for video denoising, inpainting and demosaicking that is very much related to the BM3D denoising approach [137] that we saw in Chapter 5. We recall that in BM3D each noisy patch is grouped into a 3D block with other similar patches, this block is transformed with a linear 3D transform (with a fixed basis), and the coefficients are hard-thresholded. All patches in the block are denoised simultaneously, and after the inverse transform an aggregation step performs a weighted average of all image patches that are overlapping. In [250] this procedure is done iteratively with progressively decreasing thresholds for the hard-thresholding operation; also, at each iteration the original values (at the lines that aren't missing) are restored since they are modified by the thresholding and inverse transform.

Chapter 7

White balance

7.1 Introduction

We start this chapter by recalling some basic notions of white balance that were presented in Chapter 1. The white balance process aims at emulating the color constancy ability of the human visual system and consists of two steps, illumination estimation and color correction. Ideally it should be performed not on-camera but afterwards, as offline post-processing: the rationale, as with demosaicking and denoising algorithms, is that with offline postprocessing we have much more freedom in terms of what we can do, not being limited by the constraints of on-camera signal processing (in terms of speed, algorithm complexity and so on). Offline we may use more sophisticated color constancy methods, which do not require all the simplifying assumptions used in practice for real-time processing. But offline white balance requires that the image data is stored as *raw*, i.e. as it comes from the sensor, and many cameras do not have this option, storing the image frames already in color corrected (and demosaicked, and compressed) form. If offline white balance is not an option, then the next best possibility is that of manual illumination estimation: the operator points the camera towards a reference white, such as a simple sheet of paper, and the triplet of values recorded for this object (at, say, the center of the image) is used by the camera as an estimate of the color of the illuminant. This method ensures that objects perceived as white at the scene will also appear white in the recorded images, and in general that all achromatic objects will appear gray. But cinematographers often find this effect too realistic, and use instead the manual white balance for artistic expression by fooling the camera, giving it as reference white an object with a certain color; in this way, by deliberately performing a wrong color correction, a certain artistic effect can be achieved [143]. Finally there is the option of automatic white balance (AWB), where the illumination estimation is automatically performed on-camera and the color correction (an approximation of "discounting the illuminant") is carried out on the raw data domain or just after color interpolation [334].

In this chapter we will give an overview of human and computational color constancy, with an emphasis on the working assumptions employed by the algorithms. We will end the chapter with a more detailed description of the Retinex algorithm and related methods.

7.2 Human color constancy

7.2.1 The unsolved problem of color perception

Deane B. Judd (1900-1972) was an American researcher whose work significantly advanced color science and color technology in the 20th century. He introduced colorimetry to American industry and is called "the dean of colorimetry." His writings are very precise and clear even to the uninitiated, and the interested reader is referred to the outstanding (and open-access) book [218] that collects his most influential papers. In this section we use two such works [216, 214] to pose the problem of color constancy and state the most successful scientific approaches that have dealt with it. A testament of Judd's importance in the field is that, although his papers are more than 50 years old, his findings haven't been questioned or superseded.

In [216] Judd poses the fundamental problem of color perception as the question: "Can we measure color?" and answers that while in a sense we can, in a more fundamental way we cannot. The science of colorimetry that started with Newton culminated in the CIE 1931 standard observer functions, which provide a method of measuring color that has been agreed upon by the international community. Using these functions we know that if two patches of light have the same tristimulus values (X,Y,Z) then they will look alike in color, and it's in this sense that we can measure color. This is very useful in science and technology, but the ability to specify the tristimulus values of a patch of light in a complicated, real-life scene isn't enough to predict the hue, saturation and lightness of the perceived color of such a patch. This is the fundamental, and unsolved, problem of color vision.

The color appearance of a non-emitting object depends on four factors: the spectral distribution of the illuminant, the reflectance of the object, the visual response functions of the observer, and "some as yet incompletely formulated aspect of the scene within which the object is presented" [214]. In the simple case of a flat, opaque object surrounded by a middle gray background and viewed in daylight, the color perception problem is *solved*: with the CIE 1931 standard observer functions we can predict quite accurately the hue, lightness and chroma of the object's perceived color. If one or more of these assumptions is removed, though, then the problem is *open*.

If the object is non-flat, there is an ambivalence in the observer's perception: the object appears to be of one color if it is supposed to be flat and of another if it is supposed to be solid (see Figure 7.1). Judd argues that this difference in perceptions must appear at the cortex.

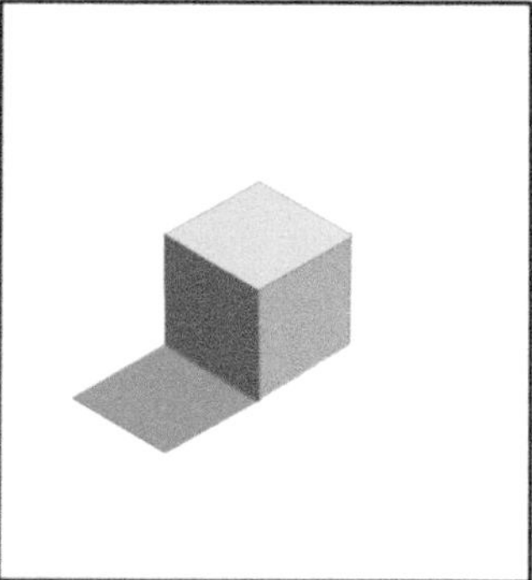

FIGURE 7.1: Left: diamond-shaped flat surfaces of different colors. Right: after arranging the surfaces into a cube, now they appear to have the same color.

If the surround is non-gray, a negative after-image of the surround color is projected onto the object's color, in what is known as *successive contrast*. If the object is next to another patch and both are seen simultaneously, the perceived color of each object is affected by the other and this is known as *simultaneous contrast*. These are phenomena that have been known for a very long time (e.g. Helmholtz mentions an account of after-images from 1634 [214]) so it has been established, at least from the mid-nineteenth century, that the perceived color of an object must depend on the colors of its neighbors and the colors previously viewed by the observer; it isn't determined just by the properties of the object itself.

If the daylight illuminant is replaced by a light source of different chromaticity, e.g. candlelight, the tristimulus values change and this difference is called *colorimetric shift*. But the observer becomes adapted, and the scene objects are seen as having (approximately) the same color as before, which is the *color constancy* property. This adaptation has at least two facets, one more rapid, on the order of seconds, the other slower, on the order of minutes.

In rapid adaptation a nonselectively reflecting object (i.e. an object with a flat reflectance function) that was heretofore perceived as gray under daylight, is now, under reddish candlelight, *seen* as reddish but still *perceived* as gray: the reddish color is associated to the ambient light and not to the object. In other words, "the observer perceives the color of the ambient light as separate from that of the object, and so discounts the nondaylight color of the ambient light" [214]. Again, this was already postulated by Helmholtz. Judd calls this process "cortical adaptation." As for how the observer perceives the ambient-light color, the most usual view in Judd's time and until the arrival of Land's Retinex theory [244] was that the illuminant is estimated as some kind of average of the colors of the scene. This is the approach taken, to the present day, by practically all computational color constancy algorithms, as we shall see in this chapter.

The slow adaptation makes the new illuminant appear achromatic after a few minutes and makes lights of other chromaticities change accordingly. Judd contended that this process takes place in the retina and calls it "retinal chromatic adaptation." Helmholtz saw the connection of this phenomenon with the three independent mechanisms of Young's theory: if an eye is adapted to candlelight the red-sensitive receptors are more stimulated and fatigued than the green and blue receptors, and the reverse happens if the eye is adapted to daylight. Therefore, any object will appear more greenish-blue under candlelight than under daylight, and this adaptive color shift can be quantified with a very simple experimental set-up where the observer has one eye adapted to daylight, the other to candlelight, and is allowed to alter the light in one of the regions so as to achieve a color match. The results of this kind of experiment were quantitatively formulated by von Kries in 1905 in what is known as *von Kries' coefficient rule*, *von Kries' rule* or *von Kries' law*: if (R, G, B) and (R', G', B') represent the colors of the same object after full adaptation to two different lights, then these values are related by the formula

$$R' = k_r R, \quad G' = k_g G, \quad B' = k_b B, \tag{7.1}$$

where the coefficients k_r, k_g, k_b are one for each type of receptor and they are independent of wavelength. The von Kries' law provides a good approximation of the experimental facts as long as the change in chromatic adaptation is moderate and the luminance of test stimulus and surround are almost constant. The coefficients can be computed by reasoning in the following way. Let (R_o, G_o, B_o) be the perceived color of the candlelight source when viewed under daylight; after adaptation, the light source becomes achromatic, therefore the values (R'_o, G'_o, B'_o) must be equal, $R'_o = G'_o = B'_o$, and we adjust k_r, k_g, k_b so that this holds.

Figure 7.2 shows the chromaticity points for 11 objects under CIE illuminant C (circles) and under a fluorescent lamp of color temperature $4500°K$ (dots). The segments going from circles to dots represent the colorimetric shift. It goes *away* from the blue region of the diagram, which is consistent with the fluorescent lamp having more energy in the blue region of the spectrum, hence adaptation to this light source makes objects look more reddish-yellow. The dots to arrow heads represent the adaptive color shift computed with von Kries' law. The distance between circle and corresponding arrow head represents the resultant color shift. We can see then that chromatic adaptation basically tries to counteract the colorimetric shift, resulting in an approximate constancy of color perception. In some cases (like in the object with Munsell notation $5P4/12$ in the figure) the color constancy is almost perfect, while in others it can be rather limited. For instance, the object $5Y8/12$ has a net shift towards green, which explains the greenish appearance given to butter by a cool-white fluorescent lamp. Worthey [382] explains how these limitations in color constancy arise from the overlap of the cone response functions, when

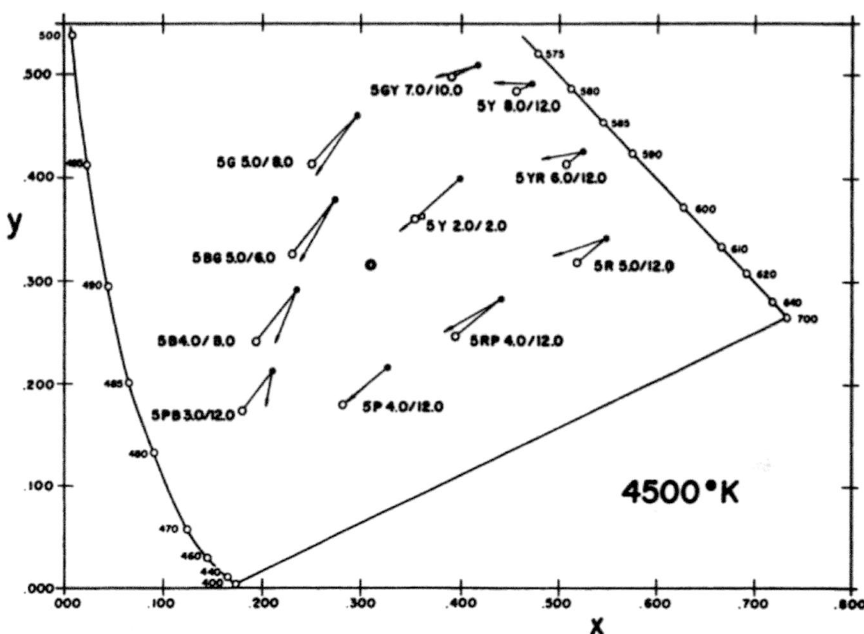

FIGURE 7.2: Chromaticity points for 11 objects under CIE illuminant C (circles) and under a fluorescent lamp of color temperature $4500°K$ (dots). Dots to arrow heads: adaptive color shift computed with von Kries' law. Figure from [214].

for von Kries' to produce perfect constancy these functions should be narrow and non-overlapping. We will return to this issue shortly.

Slow and rapid adaptation (retinal and cortical adaptation in Judd's terminology) have the same qualitative effects: by cortical adaptation we can see nonselectively reflecting objects as being gray long before the retinal adaptation makes them and the ambient light itself appear achromatic. But cortical adaptation, being fast, doesn't have the same influence as the slower retinal adaptation, with the end result that the appearance of objects changes during adaptation, especially the colors in the darkest regions of the visual field (this is known as the Helson-Judd effect).

The analytical approach to color perception starts from the chromaticity diagram and points out and evaluates the factors preventing the tristimulus values from having a correlation with color perception: general chromatic adaptation, local chromatic adaptation, discounting of illumination color and memory color. The configurational approach to color perception considers that the essence of the problem lies in the interaction of the parts of the visual field. Its basic principles are:

- Color perception works based on perceived color differences.
- Nearby elements have greater influence than distant ones.
- Any element of any scene can be replaced by a stimulus perceived as achromatic (gray or white).
- The difference between the scene element and the achromatic stimulus associated with it is what determines the color perceived at that element.

For uniformly illuminated scenes, Judd found in 1940 that the achromatic stimulus can be estimated easily, having the same chromaticity as the scene average and about half the luminance: this is precisely the "Gray world" approach to white balance and color constancy that we introduced in our chapter on on-camera processing, and pre-dates by 40 years the work of Buchsbaum [112], which is universally credited as being the first to formalize the approach. This very simple rule doesn't work on scenes with shadows, though, nor in general on scenes illuminated by more than one kind or amount of light, although artists know since the Renaissance how to overcome this problem (e.g. they know which paint to use for the part of a red object that is in shadow and which paint to use in the rest of the object, so that it appears to have a single color all over its surface).

7.2.2 Current challenges in human color constancy

In an excellent survey [175], David H. Foster identifies several key issues involved in the current understanding of human color constancy.

Measuring color constancy has the inherent complication that there might be several ways in which to judge whether or not a surface's color remains constant. In the experiments in [74], observers were presented with two identical arrays of colored papers under different illuminants, and were asked to adjust the colors of a patch so that it matched the corresponding patch in the other array by one of two criteria: either matching hue and saturation, or making the test patch "look as if it were cut from the same piece of paper as the standard patch." The results showed that "paper matches" were almost color constant, and hue-saturation matches were in the correct direction for constancy but didn't quite get there.

Quantitative measures of color constancy are based on the difference between an observer's match and an ideal value, in a given color space, either 2D (e.g. CIExy chromaticity diagram) or 3D (e.g. CIELAB). Instead of the Euclidean distance, dimensionless measures involving ratios and normalized projections are preferred, sometimes using perceptual weights.

Most experiments on color constancy are performed using simulated scenes consisting of an array of flat (and coplanar) surfaces of simple shapes. There are many practical reasons for this set-up, which allows for a simplified illumination model that ignores geometry and also permits us to discard any possible influence from familiarity or memory of the semantic content of the scene on our color constancy performance. Surprisingly, though, results from experiments with natural images show little difference from those obtained with simulated scenes: the number of objects, their shapes, their dimensionality, none of these factors seems to have a clear effect on color constancy.

A very important point to make is the following: while most color constancy algorithms in computer vision are based on an estimation of the illuminant, which is then "discounted" from the image values so as to estimate the reflectance values of the objects in the picture, there is substantial evidence that this is *not* how human color constancy works, because observers are quite insensitive to illuminant variations within a scene, estimates of illuminant may not be compatible with estimates of reflectances, very salient illumination cues are often ignored by observers, and removing cues for illumination estimation doesn't affect color matching tasks.

There is experimental evidence of three types of neural mechanisms supporting color constancy. One is cone adaptation, spatially local and consistent with von Kries' rule, providing substantial but incomplete chromatic adaptation. Another mechanism is based on spatial comparisons at the retina, the lateral geniculate nucleus, or V1: subjects receiving different stimuli at each eye have a poorer color constancy response. The third mechanism is based on the invariance of cell responses: experiments on monkeys show that neurons in several areas of the visual cortex (V1 and V4) shift their color tuning in the direction required for color constancy.

Finally, Foster [175] identifies five remaining significant problems in the understanding of human color constancy:

1. It is not clear what determines an observer's white point.
2. The relationship between color constancy and color induction needs to be clarified.
3. It is not known how surface-color attributes are combined, which of them is given perceptual prominence, nor what is their neural substrate.
4. It is not known what degree of color constancy is good enough in practice.
5. Color constancy must be reformulated, going beyond experiments with stimuli that are laboratory-based, deterministic and relatively simple.

7.3 Computational color constancy under uniform illumination

Most algorithms for color constancy work in the following way: they assume that the scene is lit by a single source, then estimate its tristimulus values and finally apply the von Kries' rule, scaling all the tristumulus values of the scene by the values of the estimated illuminant. This is precisely the method described in Section 7.2.1 to model slow chromatic adaptation in the simple case of a flat, opaque object surrounded by a middle gray background. The algorithms differ in the way they estimate the illuminant.

There are several other common assumptions, as we shall see, but the one about uniform illumination, *which allows us to apply the von Kries' rule*, is so prevalent that actually "computational color constancy" is commonly identified with this particular case, although in many situations there is more than one illuminant. This is an important point to make, especially given that we are dealing with algorithms for cinema applications, and in cinema it happens very often that scenes are lit with different light sources. In the following section we will discuss algorithms for color constancy under non-uniform illumination, all of them related to the Retinex theory of color.

The other assumptions for these color constancy algorithms is that all objects in the scene are Lambertian surfaces (diffuse surfaces, which reflect light evenly in all directions) which are also flat and parallel to the image plane [170]. In this scenario there aren't specular reflections and the geometry, location and orientation of the objects don't have influence on the amount of light they reflect, all of which are very unrealistic simplifications but such that they allow us to express the triplet of RGB values at any point in the image as:

$$
\begin{aligned}
R &= \int_{380}^{780} s_r(\lambda)I(\lambda)\rho(\lambda)d\lambda \\
G &= \int_{380}^{780} s_g(\lambda)I(\lambda)\rho(\lambda)d\lambda \\
B &= \int_{380}^{780} s_b(\lambda)I(\lambda)\rho(\lambda)d\lambda,
\end{aligned}
\tag{7.2}
$$

where $s_r(\lambda)$, $s_g(\lambda)$ and $s_b(\lambda)$ are the spectral sensitivities of the red, green and blue filters used by the camera, $I(\lambda)$ is the power distribution of the illuminant, and $\rho(\lambda)$ is the spectral reflectance of the object. In discrete terms:

$$
\begin{aligned}
R &= \sum_{i=380}^{780} s_r(\lambda_i)I(\lambda_i)\rho(\lambda_i) \\
G &= \sum_{i=380}^{780} s_g(\lambda_i)I(\lambda_i)\rho(\lambda_i) \\
B &= \sum_{i=380}^{780} s_b(\lambda_i)I(\lambda_i)\rho(\lambda_i).
\end{aligned}
\tag{7.3}
$$

We will use the overbar notation to express all the involved functions in vector form: $\bar{s}_c,\ c = r, g, b;\ \bar{I}, \bar{\rho}$. Then, Equation 7.3 arranged into matrix form becomes

$$\begin{bmatrix} R \\ G \\ B \end{bmatrix} = S'\bar{\rho}, \tag{7.4}$$

where $\bar{\rho}$ is the reflectance expressed as a column vector and S' is a 3×401 matrix with each row $c = r, g, b$ being a vector $\bar{s}'_c$ computed as the element-by-element product of a vector $\bar{s}_c$ with vector $\bar{I}$:

$$s'_c(\lambda_i) = s_c(\lambda_i) I(\lambda_i). \tag{7.5}$$

With this notation, different illuminants I originate different matrices S'. Therefore, when the same object, identified by its reflectance $\bar{\rho}$, is lighted by two different sources $\bar{I}_1, \bar{I}_2$, the camera records two different tristimuls values:

$$\begin{bmatrix} R_1 \\ G_1 \\ B_1 \end{bmatrix} = S'_1\bar{\rho}; \quad \begin{bmatrix} R_2 \\ G_2 \\ B_2 \end{bmatrix} = S'_2\bar{\rho}. \tag{7.6}$$

By the definition and properties of metamerism seen in Chapter 1, we can convert between tristimulus values of the same radiance (or of different but metameric radiances) just by multiplication by a 3×3 transform matrix. Seeing S'_1, S'_2 as two capture systems with different primaries, the two matrices S'_1, S'_2 are related by a 3×3 matrix M_{12} in which the primaries of S'_2 are expressed in terms of S'_1:

$$S'_1 = M_{12} S'_2. \tag{7.7}$$

This implies that:

$$\begin{bmatrix} R_1 \\ G_1 \\ B_1 \end{bmatrix} = M_{12} \begin{bmatrix} R_2 \\ G_2 \\ B_2 \end{bmatrix}. \tag{7.8}$$

Let $A = S'_1 {S'_2}^T$, $B = S'_2 {S'_2}^T$, where T denotes the matrix transpose and both A and B are 3×3. Then, post-multiplying both sides of Equation 7.7 by ${S'_2}^T$ we get:

$$A = M_{12} B, \tag{7.9}$$

and therefore

$$M_{12} = AB^{-1}. \tag{7.10}$$

In the case that the camera response functions have no overlap, then for each wavelength λ_i at most one value of $s_c(\lambda_i)$ $(c = r, g, b)$ is different from zero, which implies that $\bar{s}_c \cdot \bar{s}_d$ is zero if and only if $c \neq d$. Recalling Equation 7.5, it's easy to see that we get the same result for $\bar{s}'_{1c}$ and $\bar{s}'_{2d}$: $\bar{s}'_{1c} \cdot \bar{s}'_{2d} = 0$ if and only if $c \neq d$, and also $\bar{s}'_{2c} \cdot \bar{s}'_{2d} = 0$ if and only if $c \neq d$. Therefore, *if there's no overlap in the response functions* both A and B are diagonal and *the matrix M_{12} is diagonal.* But conversely, *if there is overlap in the response functions then the matrix M_{12} is not diagonal.*

Before delving into the implications of this, let's see what form M_{12} takes by computing one of its elements, say m_{11}:

$$m_{11} = \frac{\bar{s}'_{1r} \cdot \bar{s}'_{2r}}{\bar{s}'_{2r} \cdot \bar{s}'_{2r}} = \frac{\sum_i s'_{1r}(\lambda_i)s'_{2r}(\lambda_i)}{\sum_i s'_{2r}(\lambda_i)s'_{2r}(\lambda_i)} = \frac{\sum_i s_r(\lambda_i)I_1(\lambda_i)s_r(\lambda_i)I_2(\lambda_i)}{\sum_i s_r(\lambda_i)I_2(\lambda_i)s_r(\lambda_i)I_2(\lambda_i)}. \quad (7.11)$$

For a light I, the element I_R of its tristimulus value (I_R, I_G, I_B) is computed as:

$$I_R = \sum_i s_r(\lambda_i)I(\lambda_i), \quad (7.12)$$

and its energy is

$$I_R^2 = \sum_i s_r(\lambda_i)I(\lambda_i)s_r(\lambda_i)I(\lambda_i). \quad (7.13)$$

With this, Equation 7.11 becomes:

$$m_{11} = \frac{I_{1R}I_{2R}}{I_{2R}^2} = \frac{I_{1R}}{I_{2R}}. \quad (7.14)$$

Similarly, we obtain

$$m_{22} = \frac{I_{1G}}{I_{2G}}, \quad m_{33} = \frac{I_{1B}}{I_{2B}}. \quad (7.15)$$

The consequence of all this is that *if there's no overlap in the response functions* we can convert the tristimulus values obtained under one illuminant into the tristimulus values obtained under another illuminant by multiplying by a diagonal matrix whose elements are the per-channel ratios of the illuminants' tristimulus values, as we can see from Equations 7.14, 7.15, 7.10 and 7.8:

$$\begin{bmatrix} R_1 \\ G_1 \\ B_1 \end{bmatrix} = \begin{bmatrix} \frac{I_{1R}}{I_{2R}} & 0 & 0 \\ 0 & \frac{I_{1G}}{I_{2G}} & 0 \\ 0 & 0 & \frac{I_{1B}}{I_{2B}} \end{bmatrix} \begin{bmatrix} R_2 \\ G_2 \\ B_2 \end{bmatrix}, \quad (7.16)$$

which is simply the von Kries' rule:

$$R_1 = \frac{I_{1R}}{I_{2R}} R_2; \quad G_1 = \frac{I_{1G}}{I_{2G}} G_2; \quad B_1 = \frac{I_{1B}}{I_{2B}} B_2. \quad (7.17)$$

Given an image taken under illuminant I_2 we can transform it to make it appear as if was taken under illuminant I_1. Taking I_1 as a canonical illuminant, then

$$I_{1R} = I_{1G} = I_{1B} = 1, \quad (7.18)$$

and this implies two things. Firstly,

$$\begin{bmatrix} R_1 \\ G_1 \\ B_1 \end{bmatrix} = \begin{bmatrix} \frac{1}{I_{2R}} & 0 & 0 \\ 0 & \frac{1}{I_{2G}} & 0 \\ 0 & 0 & \frac{1}{I_{2B}} \end{bmatrix} \begin{bmatrix} R_2 \\ G_2 \\ B_2 \end{bmatrix}, \quad (7.19)$$

which is the formulation of the in-camera white balance process by which a color corrected triplet (R_1, G_1, B_1) is obtained from (R_2, G_2, B_2), as we saw in Chapter 3, Equation 3.7.

Secondly, from Equations 7.18 and 7.3,

$$R_1 = \sum_{i=380}^{780} s_r(\lambda_i)\rho(\lambda_i) = \rho_R$$

$$G_1 = \sum_{i=380}^{780} s_g(\lambda_i)\rho(\lambda_i) = \rho_G$$

$$B_1 = \sum_{i=380}^{780} s_b(\lambda_i)\rho(\lambda_i) = \rho_B, \tag{7.20}$$

and this shows that with Equation 7.19 we are estimating the surface reflectances ρ_R, ρ_G, ρ_B. In other words, we have shown that under all the above assumptions we can achieve *perfect color constancy.* Let us remark once more that the main assumption here is that there's no overlap in the response functions, *but this does not require that these functions be narrow-band* (i.e. sensitive at only one wavelength), as it is wrongly assumed in many works in the literature.

Human color constancy, on the other hand, is not perfect: it works better for illuminant shifts in the blue-yellow direction than for those in the red-green direction. In [382, 383] it is shown how this can be explained by the fact that there is a great deal of overlap for the medium (M) and long (L) wavelength cones, while the short (S) wavelength cones have a response that is well separated from the other two; see Figure 7.3 (left). Compensation of a blue-yellow shift requires comparison of the outputs of the S and L receptors, and since their response curves have very little overlap this comparison can be approximated well in terms of channel ratios; compensation of red-green shifts, on the other hand, requires comparison of the M and L receptors, and since their curves have a significant overlap this type of color shift will be compensated with less success. Using our notation, for human vision the matrix M_{12} is not diagonal and therefore the von Kries' rule applied in Equation 7.19 is an inexact approximation.

But Figure 7.3 (right) shows that in digital cameras the three response curves also have a significant overlap. As we explained in Chapter 3, this is due to manufacturing constraints: the response curves are given by the physical properties of the materials used for the color filter array in a single-sensor camera, and they are hard to tune at will. But this would imply that the von Kries' rule is also inexact for cameras and therefore that perfect color constancy couldn't be achieved in the computational case either. Actually, there is a work-around for this problem, which was hinted at in [383] and proposed in [168]. If S is the matrix whose rows are the camera response curves, which are overlapping, and T is another matrix of this kind but corresponding to

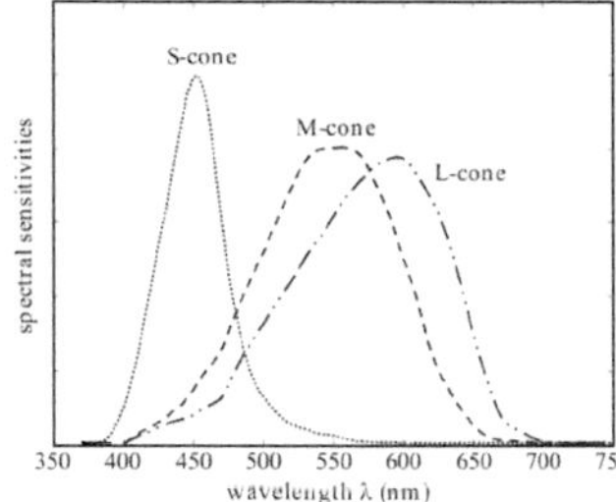

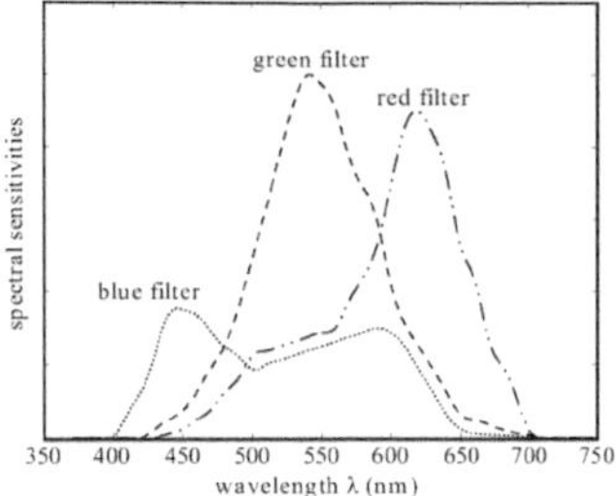

FIGURE 7.3: Spectral sensitivities of: (a) the three types of cones in a human eye, and (b) a typical digital camera. Images from [239].

non-overlapping response curves (i.e. the rows of T are orthogonal), then the camera maker can compute the 3×3 matrix W such that:

$$T = WS. \tag{7.21}$$

Keeping the same notation as before, the rows of T are $\bar{t}_c$, $c = r, g, b$. Recalling Equation 7.4 and knowing W, we can write

$$\begin{bmatrix} \hat{R} \\ \hat{G} \\ \hat{B} \end{bmatrix} = W \begin{bmatrix} R \\ G \\ B \end{bmatrix} = WS'\bar{\rho} = T'\bar{\rho}, \tag{7.22}$$

where T' is a 3×401 matrix with each row $c = r, g, b$ being a vector $\bar{t}'_c$ computed as the element-by-element product of a vector $\bar{t}_c$ with the vector $\bar{I}$ corresponding to the illuminant. Now, with two illuminants I_1 and I_2 we get:

$$\begin{bmatrix} \hat{R}_1 \\ \hat{G}_1 \\ \hat{B}_1 \end{bmatrix} = T'_1\bar{\rho}; \quad \begin{bmatrix} \hat{R}_2 \\ \hat{G}_2 \\ \hat{B}_2 \end{bmatrix} = T'_2\bar{\rho}. \tag{7.23}$$

Proceeding as before, there exists a matrix D_{12} such that $T'_1 = D_{12}T'_2$. But in this case, as the rows of T are non-overlapping, we can guarantee that D_{12} is diagonal and therefore:

$$\begin{bmatrix} \hat{R}_1 \\ \hat{G}_1 \\ \hat{B}_1 \end{bmatrix} = D_{12} \begin{bmatrix} \hat{R}_2 \\ \hat{G}_2 \\ \hat{B}_2 \end{bmatrix} = \begin{bmatrix} d_{11} & 0 & 0 \\ 0 & d_{22} & 0 \\ 0 & 0 & d_{33} \end{bmatrix} \begin{bmatrix} \hat{R}_2 \\ \hat{G}_2 \\ \hat{B}_2 \end{bmatrix}. \tag{7.24}$$

In short, the von Kries' rule can be used for computational color constancy under uniform illumination if we pre-multiply the tristimulus values by W, which in [168] was called the "sharpening transform," then apply von Kries' through multiplication by diagonal matrix D_{12}, and finally undo the sharpening transform by multiplying by W^{-1}:

$$\begin{bmatrix} R_1 \\ G_1 \\ B_1 \end{bmatrix} = W^{-1} D_{12} W \begin{bmatrix} R_2 \\ G_2 \\ B_2 \end{bmatrix}. \quad (7.25)$$

Equation 7.25 is the general model used by most if not all methods of computational color constancy under uniform illumination, which presupose that W is known and has been applied to the data. Therefore, these methods are only concerned with estimating the diagonal matrix D_{12}. Succintly put, what they do is to estimate the illuminant I_2 and then apply color correction as per Equation 7.19.

In Chapter 3 we saw the two most popular algorithms for automatic white balance, and they both belong to the class of computational color constancy under uniform illumination. They are the Gray World and the White Patch methods. In the Gray World method the illuminant is estimated as one half the average of the colors of the scene, as observed by Judd [213, 216] and formalized by Buchsbaum [112]. This method doesn't perform well if the colors of the image are not sufficiently varied (e.g. when a large, monochromatic object takes up most of the image), in which case an option would be to segment the image into differet regions, compute the average inside each one, and finally average all these averages (in this way, the size of each region does not affect the result) [150]. In the White Patch method the color of the illuminant is estimated as the color of the brightest spot in the image, which is perceived as white (the observation of this phenomenon is often attributed, incorrectly, to the Retinex theory of Land [243], but it has a long history that dates back at least to the works of Helmholtz [217, 214]). Some techniques can be used in order to prevent bright outliers from affecting the result too much: blurring the image before looking for its brightest pixel, or working with histograms [150]. Gray World and White Patch were shown to be to particular cases of a general framework based on the Minkowski norm [172].

Forsyth [174] proposed a technique for color constancy termed "Gamut Mapping" and its idea is the following. Each (strictly positive) diagonal matrix D_{12} corresponds to an estimation of the illuminant I_2, and therefore it's a possible solution to the problem. From all possible matrices D_{12}, pick that which makes the transformed image have a 3D color gamut as large as possible (within the gamut which is observable given a canonical illuminant and the sensor responses). Other options would be to choose a mean or median average of the feasible set of matrices D_{12} [169].

In [170], Finlayson et al. proposed a method called "Color by correlation" in which a set of representative illuminants is chosen, the distribution of possible colors that can be recorded with a camera under each of these illuminants is computed, and finally the illuminant is estimated as that from the original set which produces the highest correlation between the pre-computed statistics and the color distribution of the input image.

In [332], Sapiro introduced the method of "illuminant voting." It is based on ideas from the generalized probabilistic Hough transform, and the illuminant and reflectance are expressed as linear combinations of known basis functions. Each image pixel votes for possible illuminants and the estimation is based on cumulative votes.

In [107], Brainard et al. consider the color constancy problem under a Bayesian formulation. After modeling the relation among illuminants, surfaces, and photosensor responses, they construct prior distributions that describe the probability that particular illuminants and surfaces exist in the world. If the sensor response curves are known, Bayes' rule can be used to compute the posterior distribution for the illuminants and the surfaces in the scene.

Figure 7.4 compares the outputs of several AWB algorithms.

There are many other interesting approaches, like those based on using specular reflections to estimate the illuminant [360, 171], "meta-methods" that combine several different color constancy algorithms [114, 335], methods based on semantic content and memory [366], etc. The literature on color constancy for uniform illumination is really quite vast and the interested reader is referred to surveys such as [199, 150, 239, 182].

7.4 Retinex and related methods

All the color constancy algorithms described in the previous section are based on the von Kries' rule, and therefore assume that the illuminant is always global. When this isn't the case, their performance is compromised. In [268] it is shown how even small departures from perfectly uniform illumination generate considerable deviations in appearance from reflectance, which goes to say that the diagonal models of white balance, based on discounting the illumination and equating appearance with reflectance, do not predict appearances in real life scenes. As we mentioned above, this is important for us because in cinema it happens very often that scenes are lit with different light sources, which make the illumination vary locally. And furthermore, since these color constancy methods are intended to estimate reflectances rather than to simulate human perception, i.e. they strive for perfect color constancy although it's well known that human color constancy is not perfect, their suitability for cinema application is put into question.

In this section we will discuss the Retinex theory of color vision, the Retinex algorithm and related methods. The implicit assumption underlying the application of Retinex to images is that this should turn a generic input picture into a more "natural" one, i.e. an image closer to what a human observer would perceive when looking at the scene when the picture is taken [94].

7.4.1 The Retinex theory and the Retinex algorithm

Land makes in [243] a very clear and detailed explanation of his Retinex theory and the experiments that led him to its postulation. In my opinion this is Land's definitive work on the subject and probably the one that interested readers should refer to.

An experiment carried out by Land and his team consisted of arranging colored sheets of paper into two identical panels (so-called "Mondrians" because they reminded Land of paintings by Piet Mondrian, although this is quite unfair to the Dutch artist, see [186]). Quoting from [243]:
"Each 'Mondrian' is illuminated with its own set of three projector illuminators equipped with band-pass filters and independent brightness controls so that the long-wave ('red'), middle-wave ('green') and short-wave ('blue') illumination can be mixed in any desired ratio. A telescopic photometer can be pointed at any area to measure the flux, one wave-band at a time, coming to the eye from that area. [...] In a typical experiment the illuminators can be adjusted so that the white area in the Mondrian at the left and the green area (or some other area) in the Mondrian at the right are both sending the same triplet of radiant energies to the eye. Under actual viewing conditions white area continues to look white and green area continues to look green even though the eye is receiving the same flux triplet from both areas."
A variant of this experiment was a color-matching test, where the observer was presented simultaneously with the colored-sheet panel under some experimental illuminant (this was seen through thc left eye), and a standard chart of color patches, the *Munsell Book of Color*, under a "white" iluminant (seen through the right eye). In each run of the test the observer was asked to look at a given colored sheet on the left and select the Munsell patch of the right that best matched its color. The experimental illuminant was adjusted in each run so that the reference color sheet always sent the same radiance to the eye, regardless of its reflectance: gray, red, yellow, blue, and green sheets of paper were sending the same radiant energy and hence would be deemed identical by a photometer. But in each and every case, the observer was able to match the colored sheet to the closest Munsell patch, i.e. red sheet to red patch, green sheet to green patch, etc. In each matching pair, sheet and patch had different radiances but the same (scaled) *integrated reflectance*, which is defined, for each waveband, as a ratio: the integral of the radiance of the object (sheet or patch) over the waveband, divided by the integral over the same waveband of the radiance of a white object under the same illuminant. The scaling is a non-linear function that relates reflectance to lightness sensation.

Land's conclusion was that color perception had a physical correlate in these scaled integrated reflectances, which implies that in order to perceive the color of an object somehow our visual system is comparing the light coming from the object with the light coming from a reference white, since both magnitudes are needed to compute the ratio of the integrated reflectances. He wondered how we are able to find this reference white *"in the unevenly light world without reference sheets of white paper"* [243]. The sensation of white will be generated by the area of maximum radiance in all three bands (this is the von Kries' model or "white-patch" assumption, although Land doesn't cite it); this area could be used as reference, but Land didn't know how our visual system could *"ascertain the reflectance of an area without in effect placing a comparison standard next to the area"* [243]. The solution he proposed consisted of comparing far-away points through paths: the ratio of the values at the two end-points of the path can be computed as the sequential product of the ratios of each point of the path with the following point. This sequential product is reset to 1 whenever a point is reached where the value is larger than the previous maximum along the path. In this way, the sequential product for a path provides the ratio between the value at an end-point and the maximum value of the path, which is the estimate for "reference white" that can be obtained with the given path. This is the "ratio-reset" mechanism. The Retinex algorithm consists of assigning, for each point and each waveband (long, middle, short), an estimate reflectance obtained as the average of the sequential products obtained on many paths, all ending in the given point. Land thought that this was a plausible explanation of how our visual system estimates reflectances but he didn't want to venture where exactly this type of operations were being carried out, in the retina or at the cortex; therefore he chose the name "Retinex" for his approach.

It must be noted, though, that the ratio-reset idea was not a novel contribution of the original Retinex paper by Land and McCann [244], which was published in 1971. As Gilchrist points out in [183], pages 24-35: *"About 1969, a number of publications appeared that assumed the visual system to be capable of extracting luminance ratios between remote, non-adjacent surfaces, by means of a mathematical integration of chains of local edge ratios. Whittle (Whittle and Challands, 1969) was one of the first to refer to such a process."* But these works were not cited by Land, so it has become an extended misconception to believe that the edge integration idea was an original contribution of his.

The Retinex algorithm can be directly applied to digital images in a straightforward way, where the pixels will be the points and the three color channels R, G and B will be the wavebands. Over the years a great number of works have been published following and/or adapting the original formulation; see Bertalmío et al. [94] for an overview of the most relevant ones and Provenzi et al. [316] for a detailed mathematical description of the Retinex algorithm. As proved in [316], Retinex always increases brightness so it can't be directly applied to overexposed pictures; also, if the algorithm is iterated the results may improve but the convergence image is flat white, so there is some "sweet spot" of the number of iterations yielding the best output [94]. Another major source of problems is Retinex's reliance on paths: their length, shape and number condition the results and many works have been proposed trying to optimize the selection of these variables.

7.4.2 Judd's critique of Land's work

Land's Retinex has been hugely influential and also very controversial. In a scathing review of Land's work [217], Judd painstakingly shows how there isn't essentially anything new in it; furthermore, he strongly criticizes Land with biting sarcasm for not citing even one of the score of researchers who, before him, obtained the same results and made the very same observations, decades and even centuries before. In [217] Judd rejects Land's implication that *"nobody has ever before noticed that the color perceived to belong to a patch of light or an object depends on factors other than the radiant flux coming from that patch or that object"* and cites a work from 1634 to prove that *"it has long been established that a prediction of the color perceived to belong to a patch of light must be based not only on the color of that patch but also on the colors surrounding it and on the colors previously viewed by the observer."* Judd also quotes 19th century physicist Helmholtz: *"Just as we are accustomed and trained to form a judgement of colours of bodies by eliminating the different brightness of illumination by which we see them, we eliminate the color of the illumination also."* And he cites a 1938 work by Helson, who proposed to compute the ratio of the luminance of the light patches by which an object is depicted to the average luminance of the scene, in order to ascertain the perception of color. Furthermore, while Land reported as a discovery in 1959 that two-primary color systems can produce pictures where the perceived colors are of all hues, this had been a well-known and publicized fact since 1897. There are many more comments of this kind, but in a nutshell, Judd says: *"It is nevertheless remarkable that in a few years of intensive study Land, and his associates, should have been able to rediscover independently so large a fraction of the known phenomena of object-color perception."* Land's response was that *"anyone who works in the field of color deals with ideas whose intellectual genealogy is very old"* [242] and that the works cited by Judd are actually not relevant to the problem because they propose formulas that *"are wavelength-rich and time-dependent; our experiments demand formulas which are nearly independent of wavelength and fully independent of time."* While this exchange happened in 1960, somehow Land continued with his practice of not acknowledging earlier work, with the end result that today practically all the literature in computer vision and image processing attributes to Land discoveries that in most cases date back to the 18th and 19th centuries. But what is undoubtedly novel in Retinex and a key element of it is the ratio-reset mechanism presented above, which has some important implications as we will see later on in this chapter. McCann and Rizzi [269] thoroughly discuss Retinex in the context of high dynamic range imaging and point out that variational Retinex implementations that don't incorporate the ratio-reset yield results that are not consistent with the original Retinex formulation.

7.4.3 ACE

The Automatic Color Enhancement (ACE) algorithm of Rizzi et al. [328] is also based on perception, and its relationship with Retinex will become clear shortly.

ACE is designed to mimic some basic characteristics of the human visual system, like the white patch and the grey world mechanisms, lateral inhibition, the independence of chromatic channels, or the influence of spatial relationships in the scene. It consists of two stages, which are applied to each channel independently:

- In the first stage, given an input one-channel image I, an intermediate output image R is computed in this way:

$$R(p) = \frac{1}{M} \sum_j \frac{r(I(p) - I(j))}{d(p,j)}, \tag{7.26}$$

  where p, j are pixels, $r(x)$ is a non-linear, sigmoid function going from -1 when $x \to -\infty$ to $+1$ when $x \to +\infty$ and taking the value 0 at 0, and $d()$ is a distance function whose value decreases as the distance between p and j increases. The value M is a normalization constant ensuring that the maximum of R is 1, therefore R is always in the range $[-1, 1]$. This stage considers the basic principles mentioned above of chromatic independence, lateral inhibition, and spatial influence.

- In the second stage, the final output image O is computed as

$$O(p) = 255 \times (\frac{1}{2} + \frac{R(p)}{2}). \tag{7.27}$$

  This operation performs a simultaneous white patch and grey world scaling: white patch because a pixel p with maximum value $R(p) = 1$ will have an output $O(p) = 255$, and gray world because an average gray pixel p should have a value $R(p) = 0$ and therefore an output $O(p) = 127.5$.

The authors perform experiments that show how ACE has several excellent properties: it allows to obtain very good color constancy, it increases the dynamic range of the input and, unlike Retinex, it can deal both with under- and over-exposed pictures, it can perform de-quantization (eliminating quantization artifacts produced by encoding an image with an unsufficient number of bits per channel), and it can reproduce some visual perception ilusions. Its main limitation is its computational complexity, $O(N^2)$ where N is the number of pixels.

It also raises several interesting questions: what, if any, is its relationship with Retinex? If iterated, does it produce different results? And if this is the case, do they converge? We shall answer these questions in the following sections.

7.4.4 "Perceptual color correction through variational techniques"

In the work titled *"Perceptual color correction through variational techniques,"* Bertalmío, Caselles, Provenzi and Rizzi [95] start by recalling the variational histogram equalization method of Sapiro and Caselles [333], in which it is shown that the minimization of the energy functional

$$E(I) = 2\sum_x (I(x) - \frac{1}{2})^2 - \frac{1}{AB}\sum_x\sum_y |I(x) - I(y)| \tag{7.28}$$

produces an image I with a flat histogram. The range of I is $[0,1]$, x, y are pixels and A, B are the image dimensions. While histogram equalization can be performed in just "one-shot" by building a look-up table (LUT), the end result is very often unsatisfactory and can't be altered, precisely because histogram equalization with a LUT is a parameter-less and non-iterative procedure. Sapiro and Caselles propose to start with an input image I and apply to it step after step of the minimization of 7.28, letting the user decide when to stop: if the user lets the minimization run to convergence, she'll get the same result as with a LUT, but otherwise a better result can be obtained if the iterative procedure stops before the appearance of severe artifacts.

In [95] the energy in Equation 7.28 is interpreted as the difference between two positive and competing terms

$$E(I) = D(I) - C(I), \tag{7.29}$$

the first one measuring the dispersion around the average value of $\frac{1}{2}$ (as in the gray world hypothesis), the second term measuring the contrast as the sum of the absolute value of the pixel differences. But this measure of contrast is global, not local, i.e. the differences are computed regardless of the spatial locations of the pixels. This is not consistent with how we *perceive* contrast, which is in a localized manner, at each point having neighbors exert a higher influence than far-away points. Therefore, [95] proposes an adapted version of the functional 7.28 that complies with some very basic visual perception principles, namely those of white patch, locality and not excessive departure from the original data:

$$E(I) = \sum_x (I(x) - \frac{1}{2})^2 - \frac{1}{M}\sum_x\sum_y w(x,y)|I(x) - I(y)| + \lambda\sum_x (I(x) - I_0(x))^2, \tag{7.30}$$

where M is the maximum of the double-summation term, w is a distance function such that its value decreases as the distance between x and y increases, and I_0 is the original image.

In [95] it is shown that 7.30 has a single minumum and that the image I minimizing 7.30 is a fixed point of ACE. In other words, we can say that ACE is a numerical implementation of the gradient descent of 7.30, and this answers one of the questions raised in the previous section, namely, that iterating ACE we do obtain different results. Also, this iterative procedure converges (because there is a unique minimum), but for this we need $\lambda > 0$.

The minimization of 7.30 yields very good color constancy results and this method shares all the good properties and possible applications of ACE, plus the numerical implementation in [95] has a reduced complexity of $O(NlogN)$, where N is the number of pixels. Figure 7.5 compares the results obtained with this method, ACE and the Gray World AWB approach.

Apart from the color constancy application, this method can be used for contrast enhancement as well, since it produces good results without halos, spurious colors or any other kind of visual artifact.

There is a very close connection between the formulation 7.30 of [95] and Retinex, which we will present in the following section.

7.4.5 Kernel-based Retinex and the connections between Retinex, ACE and neural models

In their kernel-based Retinex (KBR) formulation [94], Bertalmío, Caselles and Provenzi take all the essential elements of the Retinex theory (channel independence, the ratio reset mechanism, local averages, non-linear correction) and propose an implementation that is intrinsically 2D, and therefore free of the issues associated with paths. The results obtained with this algorithm comply with all the expected properties of Retinex (such as performing color constancy while being unable to deal with overexposed images) but don't suffer from the usual shortcomings such as sensitivity to noise, appearance of halos, etc.

In [94] it is proven that there isn't any energy that is minimized by the iterative application of the KBR algorithm, and this fact is linked to its limitations regarding overxposed pictures. Using the analysis of contrast performed by Palma-Amestoy, Provenzi, Bertalmío and Caselles [300], the authors of [94] are able to determine how to modify the basic KBR equation so that it can also handle overexposed images, and the resulting, modified KBR equation turns out to be essentially the gradient descent of the energy 7.30 (if we also add an *attachment to data* term that leaves the image unchanged once it has departed too much from the original, as it was done in 7.30).

In this way, the connection between Retinex, ACE, and the perceptual color correction of [95] becomes explicit. Furthermore, these formulations let us make an important connection with neuroscience. As it is pointed out in [95], the activity of a population of neurons in the region $V1$ of the visual cortex evolves in time according to the Wilson-Cowan equations [108, 380, 381]. Treating $V1$ as a planar sheet of nervous tissue, the state $a(r, \phi, t)$ of a population of cells with cortical space coordenates $r \in \mathbb{R}^2$ and orientation preference $\phi \in [0, \pi)$ can be modeled with the following PDE [108]:

$$\frac{\partial a(r,\phi,t)}{\partial t} = -\alpha a(r,\phi,t) + \mu \int_0^{\pi} \int_{\mathbb{R}^2} \omega(r,\phi \| r',\phi') \sigma(a(r',\phi',t)) dr' d\phi' + h(r,\phi,t), \tag{7.31}$$

where α, μ are coupling coefficients, $h(r, \phi, t)$ is the external input (visual stimuli), $\omega(r, \phi \| r', \phi')$ is a kernel that decays with the differences $|r - r'|, |\phi - \phi'|$ and σ is a sigmoid function. If we ignore the orientation ϕ and assume that the input h is constant in time, it can be shown that Equation 7.31 is closely related to the gradient descent equation of 7.30, where neural activity a plays the role of image value I, sigmoid function σ behaves as the derivative of the absolute value function, and the visual input h is the initial image I_0. As Bertalmío and Cowan [96] point out, this suggests that the Wilson-Cowan equations are the gradient descent of a certain energy, which is a novel result, and also that there would appear to be a physical substrate at the cortex for the Retinex theory.

7.5 Cinema and colors at night

All the previous exposition in this chapter has been done assuming daytime conditions, but our perception of colors, contrast, luminance and details changes considerably in situations of low ambient illumination. In film production, night shooting on location requires more resources (more hours, extra pay, more lamps, with their trucks and generators) and it may become unfeasable because of logistics reasons (e.g. if the location can't be accessed by the trucks carrying the lamps, or if it's too large and would require an impossible amount of lighting, etc.). This is why a practice known as "Day for Night" has been common for several decades in the film industry [259], in which a scene is shot in daytime but altered (in-shoot or afterwards in post-production) to make it look like it was shot at night: the color palette is shifted towards blue, and contrast is reduced. In many cases the results are underwhelming, lacking realism, with too much blue and the same level of visual detail as in daytime. Haro et al. [193] propose a method for automatic day for night simulation that is based on emulating basic perceptual properties and psychophysical data for night vision. They take as input an image (taken in daytime) and the intended level of night-time luminance. The output is obtained through the following steps:

1. Using the standard image formation Equation 3.3 from Chapter 3, the daytime illuminant is assumed to be the standard D_{65}, reflectances are estimated from the image color values, then D_{65} is replaced by a night-time illuminant.

2. The color matching functions $\bar{x}(\lambda)$, $\bar{y}(\lambda)$, $\bar{z}(\lambda)$ are replaced by their night-time counterparts.

3. The night-time spectral luminous efficiency function $V'(\lambda)$ is obtained from tabulated data and the illumination level specified as input. This permits us to update the luminance accordingly.

4. Contrast is modified, either by emulating human contrast perception or by simulating the response of a given type of photographic film.

5. Finally, a partial differential equation (PDE) is applied to reduce details depending on the local luminance levels, emulating the loss of visual acuity in low-light situations.

Figure 7.6 shows some results obtained with this technique. Its main limitation is its assumption of a single illuminant for the whole scene.

FIGURE 7.4: Outputs of several AWB methods: (a) original image, (b) gray world, (c) white patch, (d) iterative white balancing [406], (e) illuminant voting [332] and (f) color by correlation [170]. Figure from [239].

FIGURE 7.5: Color correction results. (a) Original. (b) AWB with Gray World. (c) ACE [328]. (d) Perceptual color correction with a variational method [95]. Images (a) and (c) from [328].

FIGURE 7.6: From left to right and from top to bottom: *day for night* results of Haro et al. [193] with decreasing values of ambient illumination. Figure from [193].

Chapter 8

Image stabilization

8.1 Introduction

We saw in Chapter 3 that many cameras with CMOS sensors usually suffer from the "rolling shutter" effect: image lines are acquired sequentially, therefore there is a time difference between lines in the same frame and this causes visible distortions when there is motion in the scene: vertical motion compresses or stretches objects while horizontal motion skews them. The motion doesn't need to be particularly fast for the distortion to be significant; see Figure 8.1. CMOS sensors with an extra transistor per pixel perform global shuttering and don't suffer from this problem, but all consumer cameras and most digital cinema cameras do have a rolling shutter (the first high end digital cinema camera with a global shutter was released in late 2012 [36]).

FIGURE 8.1: The rolling shutter effect. Top: images with slowly-panning camera. Bottom: images with fast camera motion.

In hand-held camera situations the image might move too much from one frame to the next, especially when using long lenses, causing a distracting camera shake effect illustrated in Figure 8.2 (although of course this effect can only be fully appreciated when the movie is played). When shooting, this problem can be avoided with a steadicam, but this is a somewhat bulky so-

FIGURE 8.2: Frames from a video with significant camera shake.

lution and an expensive procedure that requires a special operator. Recently, small and lightweight rigs with built-in motors that provide stabilization for hand-held shots have been introduced [57], but this is still an expensive solution. Some cameras perform stabilization by displacing the lens system or the sensor in the direction oppposite to the motion, but this mechanical procedure can only handle small translational motions.

Both problems, rolling shutter and camera motion, can be dealt with in post-production using video stabilization techniques, as we shall now see.

8.2 Rolling shutter compensation

Liang et al. [252] propose a method for rolling shutter compensation that is based in two strong assumptions: that the scene is static, and the camera motion is translational and parallel to the image plane. While the first assumption is quite restrictive in practice, as Figure 8.1 shows, the second one is justified by the authors arguing that the rolling shutter effect is less noticeable with complex, non-planar motions. These assumptions allow us to model the distorted image as a transformation of the intended, undistorted image that depends solely on the relative motion between the camera and the scene. The method works in three steps:

1. Per-frame global motion estimation, yielding an initial (global) velocity estimate.

2. The velocity for each scanline is estimated after global motions are interpolated using a Bézier curve; local motion refinement is performed to improve the velocity estimate by removing outliers.

3. Finally, the image is restored by realigning the scanlines (i.e. all pixels in a scanline are displaced by the same amount).

The main shortcomings of this method stem from its assumptions. If the scene isn't static, or if the depth range of the scene is large, we will have pixels in the same scanline that should be corrected with different displacements: larger displacement for close-up pixels than for far-way pixels, or simply a displacement compensating object motion instead of camera motion. The authors suggest that these problems can be solved by decomposing the image sequence into layers and dealing with each layer separately. Also, while the algorithm handles large motion well, it isn't so effective in the case of small motion (e.g. with impulsive hand shaking), which is hard to estimate with global motion techniques.

Cho and Hong [124] propose a method that also assumes (implicitly) the scene to be static but now the motion is modeled as an affine global transformation estimated via block matching, therefore allowing for more general motions like rotations. Knowing the parameters of the affine transformation, the geometric distortions of the image can be corrected, and what is left is to remove the remaining jittering motions, which is achieved by low-pass filtering the trajectory of the 2D translational frame-to-frame motions.

Baker et al. [83] pose rolling shutter correction as a temporal super-resolution problem, where the independent motion of objects is assumed to be low-frequency and the camera jitter is assumed to be high-frequency. After optical flow computation, a high-frequency and per-scanline jitter model is estimated, which is later generalized to a high-frequency affine transformation. Independent object motion is related to the low-frequency component of the transform, which is estimated by minimizing an energy functional with a term that encourages the low frequency model to vary smoothly accross the image. These motion models are finally used to re-render each frame as if all its pixels had been captured simultaneously. The only parameter of the algorithm is the time between the capture of two subsequent rows as a fraction of the time between two subsequent frames: the larger this ratio is, the more severe the rolling shutter effects will be. This parameter can be estimated automatically from a short segment of a video containing some jitter.

Forssén and Ringaby [173] model the rolling shutter distortions as being caused by the 3D motion of the camera. They assume a pre-calibration step that yields the intrinsic camera matrix (that projects a 3D point onto the image plane), the camera frame-rate and the inter-frame delay (the time between when the last row of a frame has been read and when the first row of the following frame starts being read). This information, coupled with inter-frame correspondences obtained with a tracking procedure, allow us to parameterize the camera motion as a continuous 3D curve using calibrated projective geometry, interpolating 3D camera poses for each scanline. Finally, each row is rectified independently by a homography (a 2D perspective transformation, with respect to an estimated plane in the scene). An important limitation of this approach is the need for calibration, requiring special hardware and of course access to the camera (i.e. the method can't be applied as it is to footage from an unknown source).

Grundmann et al. [188] propose a method for rolling shutter distortion removal that, unlike [83] and [173], doesn't require prior calibration of the camera. First, they track feature points accross frames in order to perform motion estimation. These matches allow us to compute per-line homographies, which are averaged (with Gaussian weights) with the homographies of neighboring lines. The estimated homography mixtures are used to eliminate rolling shutter distortions. Camera shake is reduced by stabilizing over time similarity matrices with four degrees of freedom (2D translation, scale and rotation). At the end, a crop rectangle is applied. See some example results in Figure 8.3.

FIGURE 8.3: Removal of rolling shutter effect using the method of Grundmann et al. [188]. Original frames on the left, rectified result on the right. The model accounts for frame global distortions such as skew (left example) as well as local wobble distortions that compress and stretch different parts of the frame (right example). Figure from [188].

8.3 Compensation of camera motion

To this day, most video stabilization algorithms work in the same three stages as described by Morimoto and Chellappa [279] in 1998:

1. Motion estimation among frames, according to a certain motion model.
2. Motion compensation: unwanted camera shake is removed while preserving the intended motion of the camera.
3. Image composition: new, stabilized images are generated from the damped camera motion.

There are mainly two kinds of approaches:

- 2D methods use only 2D transformations for motion estimation and image compositions, ignoring all 3D geometry of the scene; they are fast and robust, but their quality might be limited when the scene has objects at very different depths and the motion is translational, because the induced motion parallax can't be approximated with 2D transforms.

– 3D methods use structure from motion (SFM) [194] to recover the 3D coordinates of the camera and of some feature points, and after the camera path has been smoothed new views are synthesized by reprojecting the 3D points. 3D methods may provide high quality images, but SFM is a costly and non-robust procedure that fails when the videos don't have enough information to provide 3D reconstruction ([256, 185]). For instance, this happens if the camera motion is a small translation, or the scene geometry is approximately planar. SFM is also affected when using a zoom lens, or if there is in-camera stabilization (that changes camera parameters frame by frame), or if there are large foreground objects with independent motion.

We will now briefly describe some recent methods in both categories, 2D and 3D techniques.

8.3.1 2D stabilization methods

Matsushita et al. [266] propose a method for video stabilization with two novel contributions: an inpainting technique for filling in the missing areas that appear after motion compensation, and a blind deblurring method to remove the motion blur present in the original (shaky) video, which after motion compensation becomes a distracting artifact. First they estimate global motion between frames as a homography, aligning consecutive frames through a geometric transformation. Then, local motion due to independently moving objects or to non-planar scenes is estimated through an optical flow computation, after global motion compensation. The global transformations from one frame to the next are smoothed by a Gaussian average with the neighboring transformation matrices: in this way, large and sudden displacements, which are thought to correspond to shaky camera motion, are reduced. After motion compensation, empty areas appear on the boundaries. Instead of zooming in and/or cropping, which is a common option, this method inpaints these gaps in two stages: first inpainting the motion field (optical flow) inside the gaps, then using this field to locally warp the adjacent frame so as to fill in the colors in the gaps. Finally, deblurring is performed on the motion compensated and inpainted sequence. While motion blur is (and appears) natural in the original sequence, after motion compensation we end up with a mostly still sequence that nonetheless still presents motion blur, which now appears quite out of place. Deblurring is performed by replacing blurry pixels by a weighted average of corresponding pixels from sharper frames (this procedure relies on a certain, very simple, estimation of "relative blurriness").

Lee et al. [248] propose a method based on two observations: first, that the viewer's perception of the stability of a video is often related to the motion of objects of interest, which tend to correspond to well-tracked features; second, that there usually is a wide range of possible smoothed camera motions that the viewers may accept as providing a stable enough result. From these observations, the authors introduce an algorithm that starts by extracting a set of robust feature trajectories and then finds a set of possible 2D transformations that smooth them; from these possible transformations, they choose the one that avoids leaving large blank areas after image composition. Features are found with SIFT [258], linked with neighbors using a Delaunay triangulation [106], and propagated by a feature matching procedure that also takes into account the consistency of the neighbors' motion. Each chain of linked features constitutes a trajectory; trajectories are then weighted, giving more importance to those which are longer or that come from the background or from an object of interest. Motion regularization is performed by finding a 2D similarity transform (rotation, translation and scaling) that minimizes an energy consisting of two terms: one related to the acceleration along a trajectory (penalizing motion where the velocity of features is not constant), and another term that estimates the quality of the reconstructed frames after image composition (penalizing large unfilled areas and changes of scale). Apart from the limitations inherent in 2D methods, this particular technique requires at least five trajectories through each frame for reliable results.

Grundmann et al. [189] wanted to mimic professional camera footage, so they stabilized video by making the 2D camera path be composed of segments that are either constant in time (corresponding to a static camera), have constant velocity (corresponding to a dolly shot or a panning movement), or have constant acceleration (corresponding to a smooth transition between static and moving camera shots). The original camera path is estimated by fitting 2D motion models to tracked features along the sequence. The stabilized path is found, also using a 2D motion model, as the solution of an optimization problem involving the first, second and third order derivatives of the path, but instead of L_2 minimization (where the magnitudes are squared), L_1 optimization is used (where the absolute value of the magnitudes is considered). This has the advantage that the solution will tend to satisfy the above properties regarding derivatives *exactly*, instead of on *average* as with L_2 optimization: the end result is that with this approach the optimized path will consist of segments that resemble the motion of a camera that is either static, or panning, or switching between static and panning, whereas with L_2 the result would always have some residual, shaky motion. The minimization must be subject to some constraints, otherwise the solution would just be the constant path, whose derivatives are all zero:

- the crop window must be included in the frame at all times;
- the original intended motion must be preserved, so the regularized motion can't be too far from it;
- salient points (given by a saliency map, or a face detector) should lie within the crop window.

This method is quite robust, allows for real-time results and is the one used by Youtube to stabilize clips [28]. An example result is shown in Figure 8.4. Its main limitation is that some videos can't be stabilized because the camera path estimation is unreliable (e.g. when few features can be tracked, or if most objects are non-rigid, or when there is excessive motion blur).

FIGURE 8.4: Stabilization obtained with the method of Grundmann et al. [189]. Top: original frames. Bottom: result. Figure from [189].

Liu et al. [256] propose a method that works in the following steps:

- A set of sparse 2D feature trajectories is computed.
- These trajectories (sequences of (x,y) coordinates) are arranged into a trajectory matrix.
- Each submatrix (corresponding to a few consecutive frames) of the trajectory matrix is factorized into two low-rank matrices, one of which is the eigen-trajectories matrix, representing the basis vectors that can be linearly combined to model 2D motion along this subset of frames. This is based on the work of Irani [206], who showed that for instantaneous motions the trajectory matrix (using regular, perspective cameras) is of rank 9, because the motion trajectories matrix lies on a non-linear low-dimensional manifold that can be approximated locally by a linear subspace.
- The eigen-trajectories matrix is smoothed, for instance with a low-pass filter. This creates a regularized trajectory matrix.
- Using the new feature trajectories, the original frames are warped to create the resulting, stabilized video, following the content-preserving warps approach in [255].

The authors list the following shortcomings of their method:

- in cases of dramatic camera motion, strong motion blur, geometry with little texture or large moving objects occluding the tracking geometry, not enough long trajectories are found and the method fails to produce an output;

- if a large moving object dominates the scene, a single linear subspace is not able to represent both the motion of foregound and background;
- excessive cropping may occur;
- motion blur is not removed.

Wang et al. [371] also perform stabilization by smoothing feature trajectories, but in this case trajectories are regularized by fitting Bézier curves to them, with the constrains of preserving spatial rigidity (a Delaunay triangulation is performed on the set of feature points of each frame and the resulting triangles are enforced to move rigidly) and approximating the original motion (so that the intended, low frequency motion is not lost). Finally, the frames are warped according to the regularized feature positions: each frame is represented with a regular grid mesh, and by enforcing features to be located at the smoothed positions some grid lines bend, after which pixel colors are linearly interpolated in each quad of the mesh. The limitations of this technique are very similar to those of the previous method of Liu et al. [256]: if feature tracking fails, or if foreground features are not removed, results are poor, and the method also fails if there is too much blur or not enough rigid background objects.

8.3.2 3D stabilization methods

Zhang et al. [396] propose a stabilization method where SFM [194] is used to recover the 3D camera path, which is then smoothed and finally new views are synthesized with view warping. Their feature tracking system is based on SIFT [258]. Camera motion compensation is performed by minimizing an energy with two terms: a smoothing cost that penalizes the acceleration in the rotation, translation and zoom components of the motion, and a similarity cost that penalizes departures from the original motion. View warping is performed by reprojection onto a plane with optimal constant depth, which is computed as a function of the depth range of the scene. The results of this method are heavily dependent on the accuracy of the SFM procedure, as expected.

The method by Liu et al. [255] also uses SFM to compute the 3D camera path, which is smoothed automatically via low-pass filtering or manually by fitting a user-selected path. Their contribution lies in the warping procedure, which is performed in this way:

- each frame is divided into a uniform mesh grid;
- the data term of an energy function is defined as the difference between the original and the smoothed locations of a set of feature points;
- the similarity term of the energy penalizes warps that locally (at each grid cell) don't resemble a similarity transformation; at each cell this term is weighted by the local saliency, computed as the L_2 norm of the

color variance inside the cell: in this way, points in uniform regions are allowed to be displaced more than points in textured regions;

- the sum of these two energy terms is a least-squares problem in the set of unknown grid vertices;
- the output, stabilized frames are easily generated from the warped mesh (e.g. through linear interpolation), followed by cropping.

Figure 8.5 shows an example of the results that can be achieved with this method. Its main limitations stem from the fact that it's based on SFM, which requires scenes with static regions, constantly translating cameras, a more or less uniform distribution of the feature points accross each frame, etc. Also, as in this method the stabilization can be quite aggressive, the cropping may produce a severe loss of content. Motion blur is not taken into account.

FIGURE 8.5: Stabilization obtained with the method of Liu et al. [255]. Top: final, cropped output. Middle: entire warp result. Bottom: grids and points guiding the warp. Figure from [255].

Goldstein and Fattal [185] use a mid-level 3D scene modeling method called projective reconstruction [194], with which they try to avoid the pitfalls inherent in SFM. This type of modeling recovers the geometric relationships between points and epipolar lines, not any 3D point location or 3D camera orientation, and it allows us to perform 3D stabilization without the problems of SFM-based methods. First they use a standard 2D feature tracker to estimate a few, possibly short, point trajectories. Corresponding points are used to compute the fundamental matrices relating each frame with its neighbors. With these matrices and the epipolar point transfer [162] long, virtual trajectories can be generated, then smoothed by a convolution with a Gaussian kernel. These filtered trajectories generate a new set of fundamental matrices, now corresponding to the motion of a stabilized camera, which are used to transfer every tracked point to its new location in the stabilized frame. The epipolar point transfer is also used to compute new matches in areas where the original feature tracker failed to do so, and to estimate the stabilized locations of moving objects in a way that is not affected by the static background. Finally, the new frames are generated through frame warping following the content-preserving approach of Liu et al. [255]. One of the main limitations of this method is its dependence on the number and accuracy of the tracked feature points. The method fails when there are few feature points (e.g. in scenes with little texture, or where the camera motion is excessive, or there isn't a static background), when the trajectories are very short (due to very fast camera motion), in the presence of strong occlusions or highly reflecting surfaces, or under a rolling-shutter effect. If a frame is so textured that there aren't homogeneous regions that can absorb the warping deformation, then the stabilized output will appear distorted. Finally, another limitation of this method is that it doesn't allow for the specification of a 3D camera path.

8.3.3 Eliminating motion blur

We have seen that most stabilization methods produce results that retain the original blurriness caused by motion, and the consequence is that motion blur is the most noticeable artifact in stabilized videos, as Cho et al. point out in [123]. These authors further argue that video deblurring should be performed before stabilization, not afterwards as it is customary, because most stabilization methods are based on feature tracking and this is not a reliable procedure when frames are blurry. They find that image deblurring techniques are not suited for removing motion blur in videos, for several reasons:

- blur kernels in video are introduced by camera and object motion and therefore vary both in space and time;
- deconvolution is sensitive to noise and saturated pixels;
- most multi-frame deblurring approaches rely on an accurate image alignment, and this is a very challenging problem;
- image deblurring applied on a frame-by-frame basis produces temporally incoherent results.

They propose a video deblurring method that shares the basic principles of the abovementioned work of Matsushita et al. [266] while providing more accurate results. The method is based on the observation that hand-held camera motion is highly irregular, and at the turning points of the camera motion the images are mostly sharp: these sharp pixels can be used to restore the corresponding blurry regions in other frames where the velocity of the camera motion is large. The algorithm consists of the following steps:

- Apply a standard feature tracking algorithm to compute homographies relating frames.

- Use the homographies of consecutive frames to estimate the blur function for each frame.
- Use the homographies to estimate how sharp each pixel is: when a pixel moves little from one frame to the next, according to these homographies, it is likely to be sharp.
- Divide each frame into overlapping patches, replace each patch by a weighted average of similar patches (for comparing patches, the estimated blur function is applied to sharp patches), aggregate the result by replacing each pixel by the weighted average of all corresponding pixels in the overlapping patches that include this pixel.
- Improve results by giving more weight to spatio-temporal neighbors that are sharper.
- Improve results by iteration of the whole procedure.

While this method is only capable of handling camera motion blur of static objects, the authors argue that the remaining motion blur on moving objects does not usually stand out. Some example results of this method can be seen in Figure 8.6.

FIGURE 8.6: Some sample deblurring results from [123]. Top: input frames. Bottom: after processing.

The limitations of the approach are the following. If there are large depth variations in the scene, or large moving objects, the feature tracking procedure may fail, causing in turn a wrong estimation of the blur function and of matching sharp patches in neighboring frames. When patches are too blurred, patch matching will be impossible. Saturated pixels also make the patch matching procedure fail, because they are not considered in the motion blur model used by the method. Finally, if the camera motion is always large (e.g. a constantly panning motion) then there are no sharp pixels to be used for restoration and the input is left untouched.

Chan et al. [116] propose a method for video restoration that is based in total variation deblurring [330] and that can handle motion blur, i.e. the blur kernel in each frame is spatially variant. They stack all frames in a 3D volume and operate on it, but due to the fact that the temporal resolution of videos is normally much smaller than the spatial resolution, they first need to apply a frame-rate up-conversion algorithm (see Chapter 9) to generate new frames in between the existing ones. The results are dependent on the quality of this temporal upsampling process, and the results shown look a bit oversmoothed, but probably the main limitation of the method is that it requires the spatially variant motion blur kernel to be known.

Kim et al. [229] make the following and very interesting point. The non-linear characteristic of camera response functions (CRF) makes a spatially invariant blur behave around edges as if it were spatially varying. This dampens the performance of deblurring algorithms that are based on estimating blur kernels, resulting in ringing artifacts around contours. They propose a way of estimating the CRF from one or more images and this method, while not too accurate, is enough to improve the quality of the deblurring results.

Chapter 9

Zoom-in and slow motion

9.1 Introduction

The values for the focal length of the lens and the acquisition frame rate are very important since they have a determinant influence on the final look and feel of the movie. For instance, by increasing the focal length we can zoom-in on a desired image element, and by multiplying the frame rate we can record in slow-motion. Therefore it is best to choose these values before shooting, so as to accomodate the director's vision. But it's also common that in postproduction there appears the need to perform zoom-in on a shot (e.g. because there is an element at the frame boundary that we want to get rid of, like a microphone), or to turn it into slow-motion (typically for aesthetic reasons). Of course it's out of the question to perform a re-shoot with the correct focal length and frame rate values, so we must carry out interpolation processes in space (for the zoom-in) or space and time (for slow-motion generation). Other reasons for zoom-in may be to increase the resolution of a movie (for exhibition in a higher resolution system) or a shot (to combine it with higher resolution shots). And a slow motion generation algorithm might be used for converting movies from one format to another with a different frame rate (altough this would be frame interpolation, not actual slow motion).

We will present in this chapter a brief overview of the most relevant methods for these tasks.

9.2 Zoom-in

The most simple methods for single-image zoom-in date back to the early 1980s [340] and consist of linear interpolation filtering such as bi-linear, bi-cubic or cubic spline algorithms. These are the methods that are commonly found in commercial software [176], because they are extremely fast and yield good results for small scaling factors. For large zoom-in factors, though, these techniques produce results that are visibly blurred and may suffer from artifacts around edges, such as staircasing or ringing halos.

Shan et al. [340] propose a single-image approach to image and video super-resolution. They note that single-image example-based methods such as the one of Freeman et al. [177], based in obtaining high-resolution image patches by comparison with a pre-determined image database, produce significant noise and irregularities along edges [176] and, more importantly, can't be directly extended to video because this would produce unacceptable temporal artifacts. Therefore Shan et al. don't consider high-resolution patch databases and just upsample each frame independently with a very simple technique:

1. Upsample input with bilinear interpolation.

2. Deconvolve with a non-blind deconvolution method.

3. Compute error, by downsampling the current high-resolution estimate and comparing it with the input. If the error is small then stop, otherwise continue.

4. Re-convolve and substitute pixels in the high-resolution estimate with pixels from the low-resolution input; go back to step 2.

Typically just a few iterations are needed. While the approach is fast and robust, and the authors report that no flickering is present in the resulting videos, the fact that frames are dealt with independently makes it impossible to recover fine edges or detail. This is why video super-resolution techniques don't consider frames in an isolated way and most approaches, from the mid-1980s to the late-2000s, have the following structure [139]: they register the low-resolution images to a high-resolution coordinate grid, where these images are fused to produce a high-resolution output, which is possibly de-blurred as a final step. The success of these algorithms relies heavily on the accuracy of the registration procedure. A global motion model such as a translation is not enough for most sequences, which tend to include local motion (e.g. a moving person in an otherwise static scene), but local motion models based on optical flow must be extremely accurate, otherwise the zoomed-in reconstruction will sport disturbing artifacts that make the output appear worse than the low-resolution original [315]. In fact, in their very aptly titled work "Is super-resolution with optical flow feasible?" [405], Zhao and Sawhney point out that classical super-resolution algorithms *assume* that very accurate optical flow methods exist with which the low-resolution images may be registered, but studying the requirements on the consistency of the optical flow accross images, the authors conclude that the resulting errors may actually make video zoom-in unfeasable.

Ebrahimi and Vsrcay [152] and Protter et al. [315] solved this problem by, independently and almost simultaneously, proposing essentially the same method, based on the Non-local Means (NLM) denoising paradigm of Buades et al. [109]. With their approach, the issues stemming from the one-to-one mapping of optical flow are eliminated by using the one-to-many matches and weighted average of NLM. The basic scheme of their algorithms is as follows:

- upsample all frames using any standard super-resolution technique such as bilinear interpolation: these images will therefore be blurry, an estimation of the unknown high-resolution frames after convolution with a blurring kernel, so the result must be sharpened at the end;
- replace each pixel value in each upsampled frame by the NLM weighted average obtained by considering patches in neighboring frames: this intermediate result is now a more accurate estimation of the high-resolution frame *after blurring*;
- apply a standard de-convolution algorithm such as total variation deblurring [330], and obtain the final result.

This procedure may be iterated in order to increase accuracy: in the first iteration the NLM weights are computed by comparing patches of images that have been upsampled with a simple (i.e. non accurate) method, so these weights might be a bit crude; but from the second iteration onwards, the weights can be computed directly on high-quality super-resolved frames. As a limitation, Protter et al. [315] mention the manual selection of parameters and the complexity of the method, which could be improved with a reduction of the search area. Danielyan et al. [139] use a similar approach to extend their BM3D denoising method (itself based on NLM) to the case of image and video super-resolution. Figure 9.1 shows some example results for the methods of [315] and [139].

Freedman and Fattal [176] work on a single-frame basis, as do Shan et al. [340]. They linearly upsample each frame (by a small, nondyadic factor), match patches with the input, and take from them only the high-frequency information, which they add to the upsampled image. For large scale factors they concatenate several instances of this process, and for video they just apply the method to each frame independently, reporting that no temporal artifacts are observed.

Ayvaci et al. [82] point out that recent advances in optical flow estimation make it interesting again for video zoom-in. Their approach could be seen as a combination of [152, 315] and [176] with the use of optical flow: the patch-matching of [152, 315] is restricted to neighborhoods of the optical flow correspondences, and the patch matches are used for recovering the missing high frequency information, as in [176]. Results improve on those of [152, 315], which lends credit to the assertion that there still is room for optical flow computation in video super-resolution.

Many zoom-in methods use off-the-shelf deconvolution techniques as one of the steps in their operation, and while the examples mentioned above employ the classical TV deblurring method of Rudin et al. [330], currently there are several methods that provide better results for the deblurring problem; see for instance [393, 388, 224, 353, 198].

FIGURE 9.1: Zoom-in results for the frames 1 (top), 8 (middle) and 18 (bottom) of the standard "Suzie" sequence. From left to right: low resolution input, zoomed-in just with pixel replication; original image; result of Protter et al. [315]; result of Danielyan et al. [139]. Figure from [139].

9.3 Slow-motion generation

Slow-motion generation is also known as frame rate up-conversion in the image processing literature, and it has very important applications outside cinema post-production:

- LCD displays need a higher number of frames per second than the 25 or 30 of standard content in order to reduce motion blur artifacts.
- Image compression can be enhanced by selecting some key-frames at the coding stage, discarding the rest of the frames and then re-creating these in-between frames at decoding, using a frame rate up-conversion method.
- In medical imaging, frame interpolation can be used to increase the resolution of CT scans in the $\overrightarrow{Oz}$ axis, which is quite smaller than the resolution in the xy plane.

The literature on this problem in the consumer electronics field is quite extensive, and the emphasis there is on achieving a good trade-off between visual quality and computational cost, because results must be produced in real time by the limited computational resources of a TV display. These approaches work on two stages:

- Motion estimation, where each frame is divided into square blocks of the same size and each block is assigned a motion vector that is the displacement between this block and its best match in an adjacent frame. Block matching techniques are preferred over optical flow estimation due to their lower computational cost [368].

- Motion compensation, where the new frames are interpolated along the motion trajectories determined earlier.

If motion estimation is computed from one frame to the next the method is deemed unidirectional and suffers from two kinds of problems [369]: if a pixel in the frame to be interpolated has several motion vectors passing through it there is an overlap at the pixel, and if no vector passes through it then the pixel will be empty. Bi-directional approaches estimate two vectors for each block in the frame to be interpolated, one pointing forward and the other backward. The problems of this approach lie in the difficulty of estimating the motion field and handling occlusions.

Jacobson et al. [207] and Huang and Nguyen [201] propose bidirectional approaches where the motion estimation procedure is improved by refining the motion vectors of those regions with more details, merging unreliable vectors and selecting a single motion vector for homogeneous regions.

Wang et al. [368] and Dikbas and Altunbasak [144] produce forward and backward frame interpolations that are combined with locally-varying weights depending on the reliability of the interpolation, and refined by considering a weighted average of spatio-temporal neighbors.

Liu et al. [257] and Cho et al. [125] propose methods based on tracking feature trajectories along the image sequence. In [125] each block is interpolated according to the trajectories passing through it, or those of its neighbors. In [257] several motion trajectories are used, each originating a motion compensated interpolation for each frame: these interpolations are combined according to the reliability of the corresponding trajectory. Figure 9.2 shows an example result obtained with this method.

FIGURE 9.2: Frame rate up-conversion results. From left to right: original, results of [141], [395], [403], [222], [257]. Figure from [257].

Unlike the frame rate up-conversion methods just mentioned, the following approaches aren't constrained by the needs of real-time (or even fast) performance. Takeda et al. [354] propose a 3D steering kernel regression approach, where the value for each pixel is estimated as a non-linear combination of the values of its spatio-temporal neighbors, and the non-linear weights are computed taking into account discontinuities in space and time. The method can only handle small motions, therefore a necessary pre-processing stage is to neutralize large motions with some motion compensation algorithm. This technique can be applied both for zoom-in and for slow motion generation, and the paper shows results for combined spatio-temporal upsampling, not just for frame rate upconversion. The interpolated frames appear to suffer from oversmoothing and ringing halos arround edges. Zhang et al. [397] introduce a non-local kernel regression method where each local kernel is built by considering not only its neighbors but also far away pixels (that is, this method does not rely on motion estimation) and apply it for zooming-in on videos.

Shahar et al. [339] perform frame interpolation by first generating a spatio-temporal pyramid of low resolution versions of the input video, which is treated as a 3D volume; the 3D neighborhood of each pixel in the volume is compared across the pyramid, and the best matches provide a set of constraints with which the high temporal resolution video can be generated. Salvador et al. [331] argue that the former method can only be applied to sequences with inherently repetitive motion, so that temporally aliased events can be handled effectively.

Chapter 10

Transforming the color gamut

10.1 Introduction

The color gamut of a device is the set of colors that this device is able to reproduce. Theatrical projection systems have color gamuts that are quite different from those of LCD screens, for instance, so before a movie is released it has to be processed so that its colors are tailored to the generic capabilities of each of the different display systems in which the movie is going to be shown. This process is called gamut mapping, and while it usually implies gamut compression, because the movie gamut is normally quite wider than that of common displays, the problem of gamut expansion is gaining importance as state-of-the-art digital cinema projectors are capable of reproducing gamuts surpassing even the capabilities of film.

The aim of gamut mapping is that viewers watching the same movie in different displays perceive the same colors, and it is a very challenging problem, for a number of reasons. Chief among them is the fact that color perception is a very complicated process which is far from being fully understood. And there are factors, external to the display system, such as ambient illuminance or viewer's distance from the screen, which greatly influence the final perception of colors. In practice, gamut mapping is performed in the movie industry by manually building specific look-up tables (LUT) for each movie (or specific scenes in it) and each desired output technology (HD video, digital cinema theater, etc.). A LUT is a gigantic 3D table that maps color triplets from the movie to color triplets in the display. Since each LUT has millions of entries, only a few values are manually specified by the colorist and the rest are interpolated. Still, the gamut mapping process is very much human-supervised. And for some targets, perfect color reproduction is not possible, e.g. [86] tells how the 2002 movie "The Ring" had a cyan cast in movie theaters that was completely lost and shifted to bluish-gray in TV sets because these devices can't reproduce the shades of cyan well. Also, working with LUTs imposes constraints on what can actually be achieved, because these are global transformations in color space, so all pixels with the same value are modified equally, regardless of their spatial coordinates in the image. So the colorist, while trying to color correct a scene, may find him/herself in the situation where if the colors for some objects are balanced so they look natural, this causes other objects to go off-balance and to appear unnatural. Color-balance conflicts are not uncommon when working with LUTs, and this is why the manual input of a trained colorist is so important.

In this chapter we will give an overview of the gamut mapping problem and present the state of the art in image processing techniques that perform gamut mapping in an automated way.

10.2 Color gamuts

The CIE defines color gamut this way: *"a range of colours achievable on a given colour reproduction medium (or present in an image on that medium) under a given set of viewing conditions – it is a volume in colour space."* It is important to notice then that devices do not have a unique color gamut, since there are external factors such as ambient illuminance or viewer distance that influence the way colors produced by the device are perceived. Although this definition of color gamut applies only to media where outputs are/produce color stimuli (e.g. printers, displays, projectors), there is work also on computing gamuts for input media (camera, scanners) [285].

Different media and different display technologies have different color gamuts, as Figure 10.1 shows. For instance, the color gamut of film is quite wider than that of a CRT TV set, so a direct input of the film color values to the CRT device would reproduce incorrectly all colors outside the CRT's gamut, producing visible color artefacts, as we discussed in Chapter 3. Therefore, it is necessary to establish correspondences between colors in both gamuts, a mapping in color space, and this is the gamut mapping (GM) problem. Most gamut mapping algorithms (GMAs) require the gamuts to be described in terms of their boundaries, with a gamut boundary descriptor (GBD).

The gamut boundary of a device may be characterized either by analytical methods that model the device's properties, or by measuring the positions in color space of a sparse set of test colors. In the latter case, the GBD has to be estimated from the experimental pointset and there are many algorithms that do this: in [282], Morovic highlights the methods of computing the convex hull of the point set [236], doing a sort of morphological closure (*alpha shapes*, [127]), and splitting color space into spherical segments and computing the extrema for each segment (*segment maxima*, [283]).

Usually the source gamut is wider than the destination gamut, and the GM operation that has to be performed is gamut reduction. Otherwise, the process is called gamut expansion or gamut extension. There is extensive literature on GMAs; see [284] for a survey on early works and [282] for a more recent and very detailed account. As pointed out in [105], GM algorithms originally were non-adaptive, just performing gamut compression or gamut clipping in order

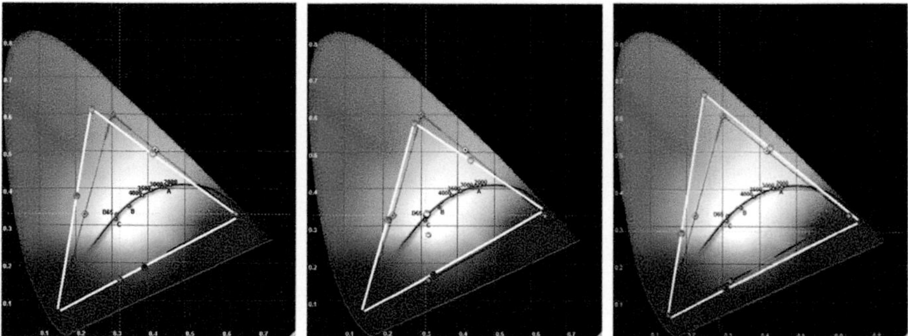

FIGURE 10.1: The color gamut (as white triangles) of different TV display technologies, with reference to the BT.709 color space (black triangle). From left to right: CRT, LCD, plasma. Images by P.H. Putman [317].

to fit the color gamut of an image into the gamut of a given device. Next, GM algorithms that adapted to image type and image gamut were introduced, and currently some of the best performing GM methods are adaptive to the *spatial* content of the image, which is something that can't be achieved by operating on LUTs. But the GM problem is far from being solved, usually an algorithm that performs well for a pair of gamuts will not give good results for other gamuts or even other images within those gamuts, to which there is the added difficulty of evaluating the performance of GMAs, as we discuss in a later section.

10.3 Gamut reduction

The simplest gamut reduction technique is gamut clipping, in which colors inside the destination gamut are left untouched while those lying outside are "clipped," projected onto the gamut boundary. Figure 10.2 shows different approaches:

- straight clipping retains the luminance and substitutes the chroma for that of the closest color in the boundary;
- node clipping maps to gamut boundary colors in the direction to the mid-gray value ($L* = 50$);
- cusp clipping maps to gamut boundary colors in the direction to the point in the achromatic axis (the $L*$ axis) corresponding to the maximum $C*$ value at the given $h*$ plane;

– closest $L*a*b*$ clipping maps to the closest point in the gamut boundary.

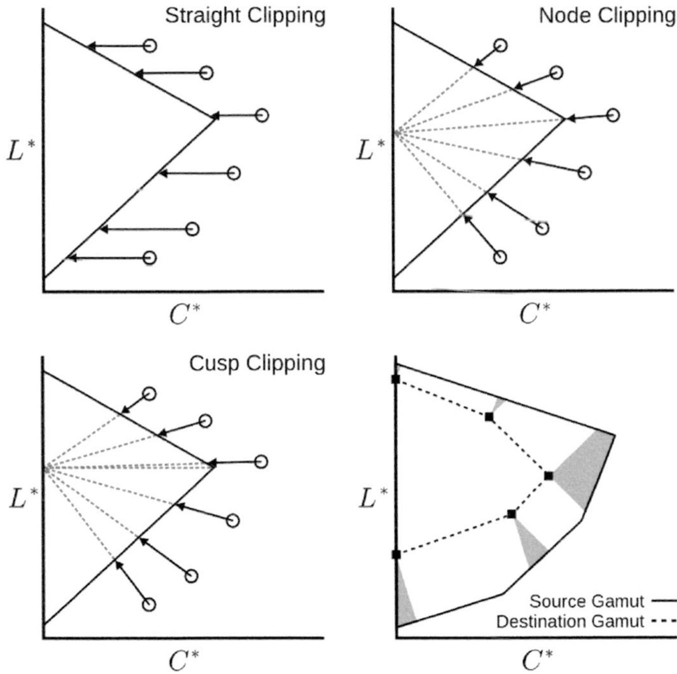

FIGURE 10.2: Different gamut clipping techniques, figures adapted from [278, 282].

All gamut clipping techniques map whole segments into single points, and this may produce visible loss of detail in the mapped result as well as the appearance of color gradients.

To avoid this sort of problem, gamut compression algorithms operate over all colors, both inside and outside the destination gamut. The majority of GMAs fall in this category, and there is a great diversity of approaches. Apart from *"compression"* versions of the clipping algorithms mentioned above, where now segments are mapped into segments instead of points on the boundary, the possibilities and variations are very many:

- the compression itself may be performed in several ways (linear, different non-linear alternatives);
- the mapping may proceed sequentally dimension after dimension (e.g. lightness scaling followed by chroma modification) or it may deal with all dimensions at the same time;
- the "focal point" towards which the mapping is realized may be one or there may be several, in one or several axes;

- some subregion of the gamut may be left untouched, as clipping algorithms do;
- different mapping techniques may be used in different gamut regions;
- the color set of the input image may be used as source gamut instead of the color gamut of the source medium.

All gamut clipping and gamut compression GMAs operate on colors, regardless of the spatial position of the pixels that have those colors. This may be problematic and cause a visible loss of detail. By mapping a line segment in color space into a shorter segment (or into a point, as in the case of gamut clipping), all pixel values within the source segment get mapped into a smaller set of colors: this does not usually matter if the pixels are far apart, but if they are neighbors, it may be the case that they were distinguishable in the source image but after the mapping they look identical in the destination (e.g. if their new colors fall in the same color quantization bin).

Spatial gamut reduction algorithms take into account both the color gamuts and the spatial distribution of colors. We may categorize spatial GMAs into:

- Frequency-based approaches, which perform low-pass or band-pass filtering on the source image (on either its luminance or the three color channels), apply a clipping or compression GMA on this modified source, and then add back the high frequency detail to the final result. A pioneering work in this area was that of Meyer and Barth [274]. Many variations have been proposed, like filtering the difference image instead of the source, employing edge-preserving bilateral filtering [407], or using a multi-scale approach to perform GM sequentially from the bottom to the top of the scale pyramid [286].
- Ratio-preservation approaches, like [267, 230, 160, 72], which try to maintain the image gradient when performing the mapping. This is rooted in the principles enunciated in the Retinex theory of color, that state that perception is related to comparisons or ratios of image values, rather than to the actual values themselves, as we saw in Chapter 7.
- Optimization approaches, like [293, 230], that use a perceptual metric to compute the difference between source and destination, and minimize this metric in order to find the desired in-gamut result.

10.4 Gamut extension

A television signal has a limited color gamut, referred to as EBU (European Broadcasting Union) or Rec. 709. This EBU gamut is wider than that of

CRT displays, so it has been adequate for many years, but newer display technologies that use more than three primaries or where the primaries are more saturated are capable of achieving a gamut that is wider than EBU's [238], e.g. see Figure 10.3. In this case, the gamut mapping operation required is not compression but extension.

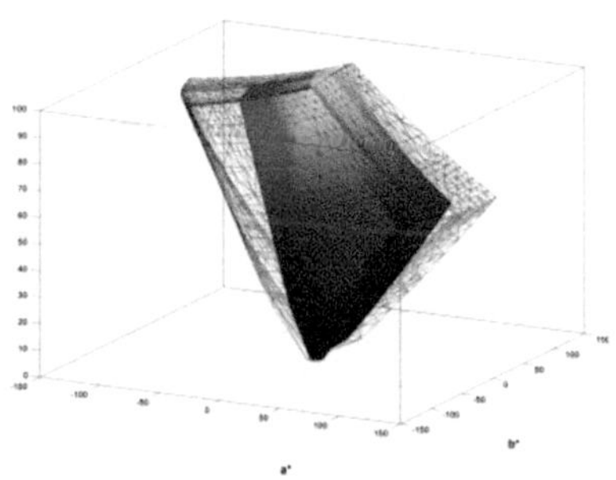

FIGURE 10.3: Television gamut (in color, inside) and the gamut of a state-of-the-art TV display (outside, wireframe) [238].

Another example would be in digital cinema projection: state-of-the-art digital projectors have a very wide gamut, but often the cinema signal inputs they receive have been coded with a limited gamut as a precaution against commonly poor projection systems. The end result is that state-of-the-art projectors do not realize their full potential in terms of color rendition.

While it is true that there is considerably less literature on gamut extension than on gamut reduction, one could think of simply taking any gamut compression technique and use the one-to-one mapping in the reverse direction, as Morovic comments in [282]. Kim et al. [227] proposed an extension-specific GMA with three types of extension: in chroma (constant lightness), using the origin as focus (which they called *vector*) and with a variable focus, depending on lightness (which they called *adaptive*); see Figure 10.4.

Previously, Kang et al. [220] had proposed a gamut extension method of lightness and chroma mapping that was derived as the numerical fitting of subjective data derived from user tests. Laird et al. [238] propose and test several gamut extension algorithms, inspired by [227]. While they report that their test users tended to prefer lightness-chroma mapping (based on Kim et al.'s *adaptive* method), they also point out that the performance of all algorithms changed notably among images, and that there often were unnatural colors in the resulting images.

Finally, none of these gamut extension methods is space-dependent, which suggests a possible route for improvements in this particular problem.

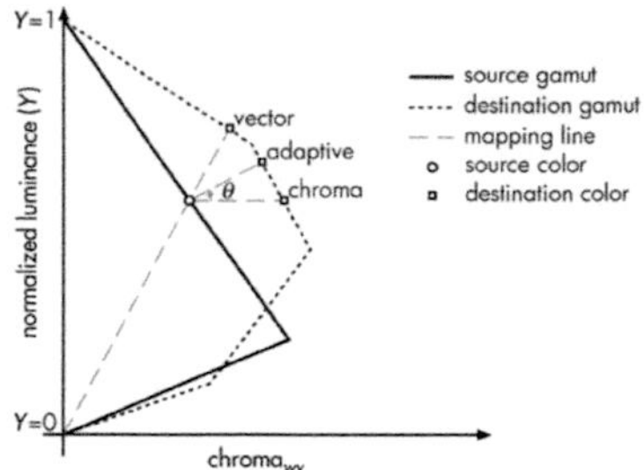

FIGURE 10.4: Kim et al.'s extension algorithms [227]. Figure taken from [282].

10.5 Validating a gamut mapping algorithm

As Morovic [282] very elegantly says, *"the vast majority of algorithms have been either evaluated only by their authors and (almost without exception) been found to outperform other algorithms that they were compared with, or they were evaluated by more than one group of researchers, who as a result of differences in the evaluation conditions have found contradictory results."*

In order to validate GM algorithms, perceptual studies are conducted following the recommendations of the CIE [357], and this is a rather involved process, requiring a windowless room with neutral grey walls to perform the experiments in, regulated ambient light and specialized and calibrated equipment. The recommendations also cover many other aspects like image test set, gamut boundary computation, color spaces, etc. The most popular evaluation method is pair comparison, where each observer has to choose which of two different gamut-mapped versions of an image is more faithful to the original. This allows us to build a ranking of GM algorithms, but it being a complicated and subjective procedure it is not surprising that results differ greatly among studies. For instance, while in [105] spatial GMAs are found to surpass color-by-color GMAs for almost every image, in [147] one of the best performing algorithms is not a spatial GMA. Interestingly, [147] also reports that non-expert observers could not really distinguish among different GMAs, which really puts things into perspective. Another shortcoming is that this type of evaluation does not offer clues as to how to improve any given GM method. Some objective, quantitative quality metrics have been proposed that are inspired by knowledge of the human visual system [157, 372, 401], but several works report that there is no correlation between results obtained with said metrics and with perceptual evaluation experiments [105], therefore "it isn't currently possible to evaluate color gamut mapping quality using color image difference metrics" [192]. Recently, a perceptually based color image difference metric [254] has been proposed that particularly emphasizes the assessment of gamut-mapped images. It is based on predicting the distortions in lightness, hue, chroma, contrast and structure of the gamut-mapped images by performing the comparison with the original images.

10.6 An example of a spatial gamut reduction algorithm

In this section we present in some detail a recent spatial gamut reduction algorithm by Zamir, Vazquez-Corral and Bertalmío [394], reproducing excerpts and figures from that paper.

10.6.1 Image energy functional

As we saw in Chapter 7, Bertalmío, Caselles, Provenzi and Rizzi proposed an image enhancement model in [95] where the image energy functional is defined as

$$E(I) = \frac{\alpha}{2}\int\limits_{\mathfrak{I}}\left(I(x)-\frac{1}{2}\right)^2 dx + \frac{\beta}{2}\int\limits_{\mathfrak{I}}\left(I(x)-I_0(x)\right)^2 dx \\ - \frac{1}{2}\int\int\limits_{\mathfrak{I}^2} w(x,y)|I(x)-I(y)|dxdy, \tag{10.1}$$

where $\alpha \geq 0, \beta > 0$, I is a color channel $(R, G$ or $B)$, $w(x,y)$ is a normalized Gaussian kernel of standard deviation σ, and $I(x)$ and $I(y)$ are two intensity levels at pixel locations x and y respectively.

This functional has two competing parts. The positive competing terms are global: the first one controls the dispersion from the middle gray value, which in images in the range $[0,1]$ is assumed to be $1/2$ as described in the gray-world hypothesis [112], whereas the second term in the functional penalizes the departure from the original image I_0. The negative competing term represents the local contrast. Therefore, by minimizing the image energy $E(I)$, the aim is to maximize the contrast, while not departing too much from the original image, and to preserve the gray-world hypothesis. It is formulated in [95] that the steady state of the energy $E(I)$ can be achieved using the evolution equation

$$I^{k+1}(x) = \frac{I^k(x) + \Delta t\left(\frac{\alpha}{2} + \beta I_0(x) + \frac{1}{2}R_{I^k}(x)\right)}{1+\Delta t(\alpha+\beta)} \tag{10.2}$$

where the initial condition is $I^{k=0}(x) = I_0(x)$. The function $R_{I^k}(x)$ indicates the contrast function:

$$R_{I^k}(x) = \frac{\sum_{y \in \Im} w(x,y) s_m \left(I^k(x) - I^k(y) \right))}{\sum_{y \in \Im} w(x,y)} \tag{10.3}$$

where x is a fixed image pixel and y varies across the image. We define the slope function s_m for a real constant $m > 1$, when $d = I^k(x) - I^k(y)$, as follows

$$s_m(d) = \begin{cases} -1, & \text{if } -1 \leq d \leq -\frac{1}{m} \\ m \cdot d, & \text{if } -\frac{1}{m} < d < \frac{1}{m} \\ +1, & \text{if } \frac{1}{m} \leq d \leq 1 \end{cases} \tag{10.4}$$

Bertalmío, Provenzi and Caselles show in [94] that the presented model is related to the Retinex theory of color vision proposed by Land [244], and that the model can be adapted to perform contrast reduction by just changing the sign of the contrast term. In the next section we will explain how we can use this contrast reduction property for gamut mapping.

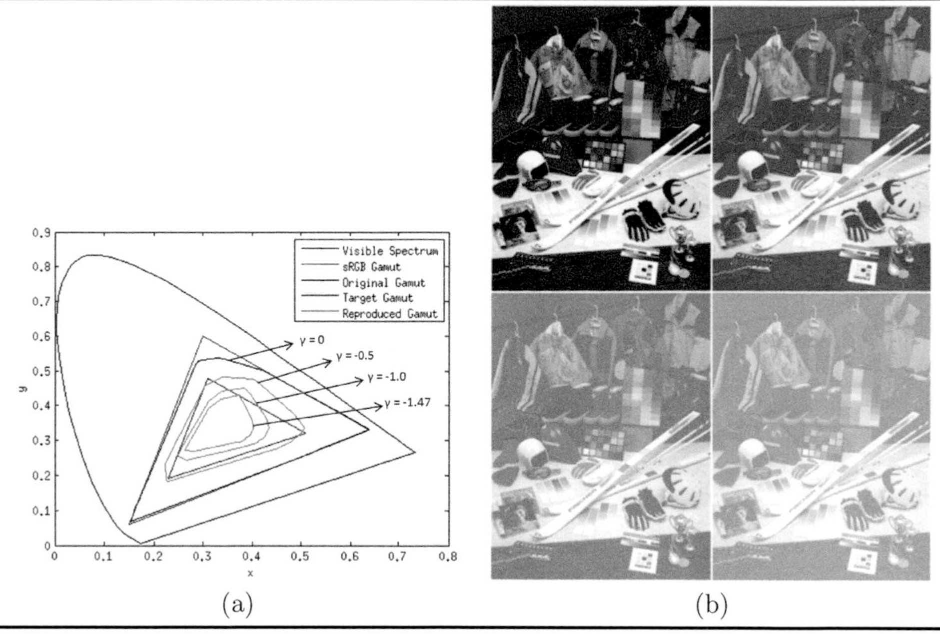

FIGURE 10.5: Perceptual GM Approach. (a): Gamuts on chromaticity diagram. (b): Top left: original image. Top right: $\gamma = -0.50$. Bottom left: $\gamma = -1.0$. Bottom right: $\gamma = -1.47$. Figure from [394].

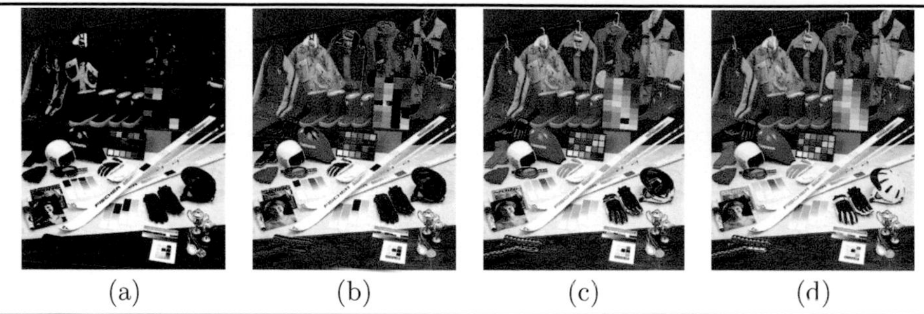

FIGURE 10.6: Gradual mapping of colors. Out-of-gamut colors (in black) when (a): $\gamma = 0$, (b): $\gamma = -0.50$, (c): $\gamma = -1.0$, (d): $\gamma = -1.47$. Figure from [394].

10.6.2 Gamut mapping framework

In this section, we adapt the image energy functional defined in Equation (10.1) to perform gamut mapping from the gamut of the original image to the target gamut of a given device. In order to control the strength of the contrast modification, we add the contrast coefficient γ in the image energy functional $E(I)$. Recall that α controls the dispersion around middle gray. Since, in the case of gamut mapping, the human visual system adapts to the luminance of the environment instead of the luminance of the stimulus, we set $\alpha = 0$ and the image energy model defined in Equation (10.1) becomes

$$E(I) = \frac{\beta}{2} \int_{\mathfrak{I}} (I(x) - I_0(x))^2 \, dx - \frac{\gamma}{2} \int\int_{\mathfrak{I}^2} w(x,y)|I(x) - I(y)| dx dy. \quad (10.5)$$

The evolution equation (10.2) reduces to

$$I^{k+1}(x) = \frac{I^k(x) + \Delta t \left(\beta I_0(x) + \frac{\gamma}{2} R_{I^k}(x)\right)}{1 + \beta \Delta t} \quad (10.6)$$

where $\gamma \in \mathbb{R}$ is positive or negative depending on whether we want to maximize or minimize the contrast, respectively [94]. In our case, γ will always be negative, since our goal is to reduce the contrast in order to perform gamut reduction.

The evolution equation (10.6) has a steady state for each particular set of values for $\beta, \Delta t$ and γ. For example, in Figure 10.5 (a), a chromaticity diagram is shown with different gamuts (visible spectrum, sRGB gamut, original gamut, target gamut and reproduced gamut). It can be shown that when $\gamma = 0$ the steady state of the evolution equation is equivalent to the original image. In the same figure we show that as γ decreases, the steady state of Equation (10.6) has a gamut that is gradually smaller. Figure 10.5 (b) shows that, just

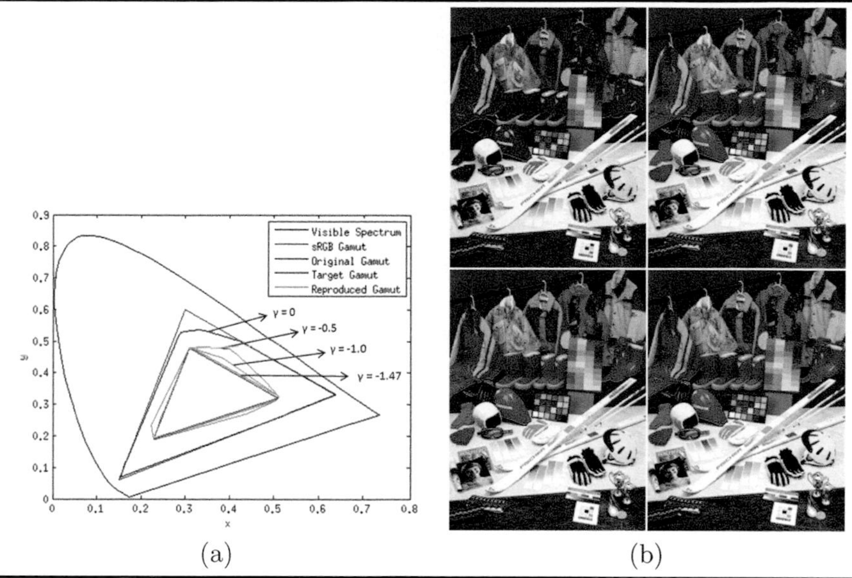

FIGURE 10.7: Modified Perceptual GM Approach. (a): Gamuts on chromaticity diagram. (b): Top left: original image. Top right: $\gamma = -0.50$. Bottom left: $\gamma = -1.0$. Bottom right: $\gamma = -1.47$. Figure from [394].

by selecting a value for gamma that is small enough ($\gamma = -1.47$ in this case) we are already performing a gamut mapping algorithm. However, now colors that were originally inside the target gamut move inwards too much, and the image becomes washed-up, as the figure shows.

In order to improve the previous result, we present an iterative method in terms of the contrast coefficient γ. At iteration 1, we set $\beta = 1$ and $\gamma = 0$, and therefore the original image is obtained as the steady state. We leave untouched the pixels that are inside the destination gamut, and we move to iteration 2, where we decrease γ (for example, setting $\gamma = -0.05$) and run Equation (10.6) to steady state. In this second iteration, we check whether any of the points that were outside the gamut at the previous iteration have been moved inside the destination gamut. If this is the case, we leave them untouched for the following iterations. We keep iterating by decreasing γ until all the out-of-gamut colors come inside the destination gamut. An example of this iterative procedure is shown in Figure 10.6, where black pixels represent out-of-gamut pixels left in that iteration. It can be seen in Figure 10.7 (a) that the reproduced gamut is covering a much wider range of colors than previously. It is shown in Figure 10.7 (b) that the colors are better preserved as compared to the previous example (see Figure 10.5).

FIGURE 10.8: Effect of standard deviation (σ). Top left: original. Top right: $\sigma = 25$. Bottom left: $\sigma = 100$. Bottom right: $\sigma = 200$. Figure from [394].

10.6.3 Experiments

We work in the RGB domain by fixing the parameters $\beta = 1, \Delta t = 0.10$ and iterate by decreasing the parameter γ ($\gamma \leq 0$) until the colors of the original image come inside the target gamut. For each value of γ we run Equation (10.6) to steady state, which we assume has been reached when the difference between two consecutive steps falls below 0.5%. We have noticed that the standard deviation σ of the Gaussian kernel w is of great importance; we observe in Figure 10.8 that a small value of σ leads to the preservation of colors but introduces artifacts, whereas for the larger values of σ each color pixel is strongly influenced from the surrounding colors. Therefore, we compute the gamut mapped images $\mathcal{I}_\sigma$ by using four different values of standard deviations $\sigma \in \{50, 100, 150, 200\}$. Subsequently, in order to obtain a final gamut mapped image $\mathcal{I}_{final}$, we combine all the outcomes $\mathcal{I}_\sigma$ with respect to the original image $\mathcal{I}_{orig}$, in Lab color space, by using the ΔE measure.

$$\mathcal{I}_{final}(x) = \arg\min_{\mathcal{I}_\sigma} \left(Lab(\mathcal{I}_\sigma(x)) - Lab(\mathcal{I}_{orig}(x))\right)^2, \qquad \sigma \in \{50, 100, 150, 200\} \tag{10.7}$$

We are confident that varying these parameters according to the application and image characteristics would give better results. However, our choice of parameters is the same for all the results shown in this paper.

FIGURE 10.9: Gamut mapping results. Column 1: original images. Column 2: output of HPMINDE clipping [289]. Column 3: output of Lau et al. [245]. Column 4: output of Aslam et al. [73]. Column 5: output of our algorithm. Figure from [394].

10.6.3.1 Qualitative results

In this section, we apply our method on a rather challenging target gamut as shown in Figure 10.5. Given an image in sRGB, our algorithm maps the gamut of the original image into the destination gamut. The results presented in Figure 10.9 show that our proposed framework works well in preserving the colors, texture and color gradients from the out-of-gamut regions while staying faithful to the perception of the original image. For example, in Figure 10.10, rows 1 and 4, it can be seen that the colors reproduced by our GM algorithm (fifth column) are much more saturated than those of HPMINDE [289] (second column), and the state-of-the-art algorithms of Lau et al. [245] (third column) and Aslam et al. [73] (fourth column). Similarly, in Figure 10.10, row 2, our algorithm not only reproduces the color efficiently but also preserves a great amount of texture. In Figure 10.10, row 3, we can see our method accurately represents the difference in the lightness of identical hue (see the pink socks and pink beanie). Results show that our algorithm outperforms not only the widespread HPMINDE method [289] but also the state-of-the-art algorithms [73], [245].

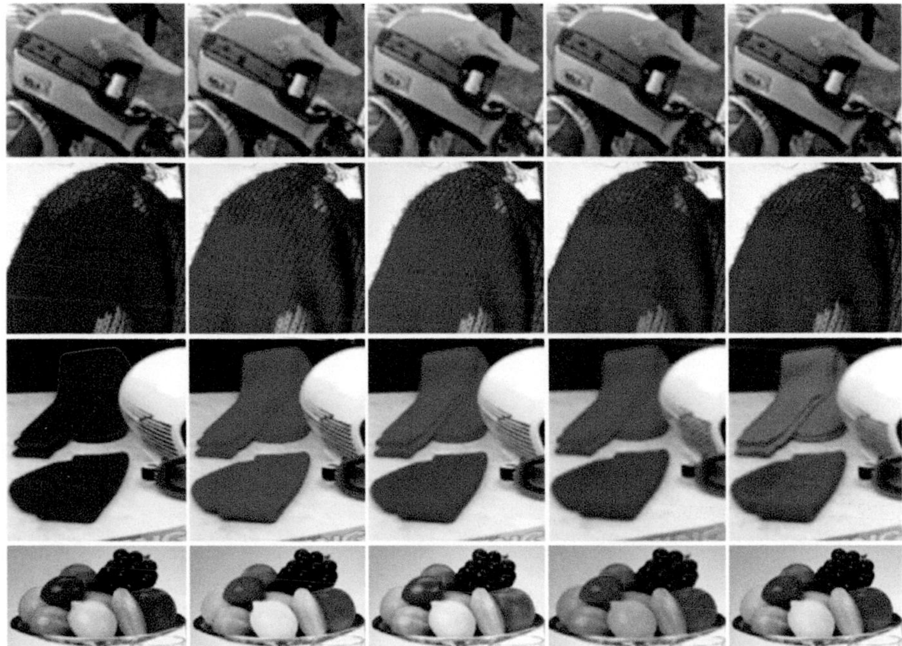

FIGURE 10.10: Preserving details, all images are cropped from Figure 10.9. Column 1: original cropped images. Column 2: output of HPMINDE [289]. Column 3: output of Lau et al. [245]. Column 4: output of Aslam et al. [73]. Column 5: output of our algorithm. Figure from [394].

10.6.3.2 Objective quality assessment

Visually, the results presented so far underline the good performance of our GMA in terms of visual quality. This subjective outcome is backed by using the perceptual color quality measure presented in [254]: the Color Image Difference (CID) metric estimates the perceptual differences given by the changes, from one image to the other, in features such as hue, lightness, chroma, contrast and structure.

Comparisons using the CID metric are provided in Table 10.1. In this table we can see that our algorithm outperforms the other methods in 15 out of 17 test images. Moreover, the statistical data (mean, median and root mean square) is also presented in Table 10.2. These results show that our method produces a gamut mapped image which is, perceptually, more faithful to the original image as compared with the other methods.

TABLE 10.1: Quality assessment: perceptual difference measure [254]. Method 1: HPMINDE [289]. Method 2: Lau et al. [245]. Method 3: Aslam et al. [73]. Table from [394].

	Method 1	Method 2	Method 3	Proposed [394]
Caps	0.1027	0.1022	0.0821	**0.0711**
Raft	0.0772	0.0747	0.0857	**0.0471**
Barn	0.0268	0.0242	0.0134	**0.0088**
Girl	0.0825	0.0695	0.0359	**0.0209**
Birds	0.1829	0.1119	**0.0923**	0.1086
Bikes	0.0330	0.0396	0.0322	**0.0155**
Boat	0.0255	0.0187	0.0035	**0.0008**
Beach	0.0168	0.0151	0.0077	**0.0046**
Party	0.0569	0.0878	0.0487	**0.0280**
Portrait	0.0235	0.0393	0.0209	**0.0104**
Picnic	0.0954	0.0954	**0.0448**	0.0638
Window	0.0514	0.0591	0.0443	**0.0326**
Woman	0.1313	0.0882	0.0528	**0.0410**
Boats	0.0183	0.0287	0.0195	**0.0130**
Statue	0.0025	0.0061	0.0053	**0.0020**
Model	0.0292	0.0736	0.0398	**0.0390**
Ski	0.1899	0.1964	0.1734	**0.1040**

TABLE 10.2: Quality assessment: statistical data. Table from [394].

	Mean	Median	RMS
HPMINDE [289]	0.0674	0.0514	0.0873
Lau et al. [245]	0.0665	0.0695	0.0807
Aslam et al. [73]	0.0472	0.0398	0.0627
Proposed [394]	**0.0360**	**0.0280**	**0.0485**

10.7 Final remarks

Gamut mapping remains a challenging open problem of interest to the movie industry for the foreseeable future.

Despite the continuous evolution of display technology, it is unavoidable that older, smaller-gamut displays remain a significant part of all the displays used for quite some time, therefore justifying research into gamut reduction algorithms. But even if most displays were wide-gamut, there is a tremendous amount of legacy video content in reduced gamuts such as Rec. 709, for which gamut extension algorithms are needed.

While several works suggest that image-based, spatial GMAs are the ones that perform best, it is important to remark that one of the main challenges in devising GMAs comes from the fact that these algorithms are very hard to evaluate, and hence to improve. User test studies are quite complicated and usually give contradictory results, and quantitative tests based on perceptual metrics may not correlate with user experiences. It seems paramount then to work in GMA evaluation in order to get closer to solving the GM problem.

Chapter 11

High dynamic range video and tone mapping

11.1 Introduction

The ratio between the brightest and the darkest values that a camera can capture is called dynamic range, and most cameras take low dynamic range pictures, spanning only a few orders of magnitude. As a consequence, if the light coming from a scene to the camera is of high dynamic range, then using a short exposure time, the camera will render correctly the details in the brightest parts of the image, but it won't be able to capture details in the dark regions. Conversely, using long exposures we are able to see the dark regions but the bright ones become flat white (overexposed or "burnt"). And there is no exposure value that works for both types of regions; see Figure 11.1 for an example.

But the dynamic range of our visual system, while also limited, is 1-2 orders of magnitude larger than that of cameras and therefore there are many situations in which we are able to perceive with the naked eye details and contrast that a camera is not able to capture unless some artificial lights (lamps in cinematography or professional photography, a flash in amateur photography) are added to the scene. The effect of these extra lights is to increase the brightness of the dark regions, thus reducing the ratio between brightest and darkest value of the scene, i.e. its dynamic range. This is an issue of capital importance in cinema production. A large percentage of the shooting time in any movie is devoted to lighting the scenes. And commonly much of this time is not dedicated to lighting for mood and artistic expression but to *basic lighting*, i.e. adding extra lights to the scene so that it looks, in the movie, very similar to how it looks *without extra lights* to an observer present in the scene. Hence it is normally the camera system that requires the extra lighting, not the people present at the shooting; see Figure 11.2.

The great director Sydney Lumet clearly explains this in [259] (page 83):

If you've ever passed a movie company shooting on the streets, you may have seen an enormous lamp pouring its light onto an actor's face. We call it an arc or a brute, and it gives off the equivalent of 12,000 watts. Your reaction has probably been: What's the matter with these people? The sun's

FIGURE 11.1: Pictures taken with different exposure settings in daylight. Each picture is good only for either the foreground (trunk) or the background (trees, sky) details **but no exposure allows us to capture both**. A human observer is in fact able to perceive details and color on the foreground and background **simultaneously**. In cinema or professional photography, this would be achieved by setting up a lamp to shed light on the foreground so that it becomes brighter; in amateur photography we would use the camera's built-in flash. Source image by ILM, figure from [164].

shining brightly and they're adding that big light so that the actor is practically squinting. Well, film is limited in many ways. It's a chemical process, and one of its limitations is the amount of contrast it can take. It can adjust to a lot of light or a little bit of light. But it can't take a lot of light and a little bit of light in the same frame. It's a poorer version of your own eyesight. I'm sure you've seen a person standing against a window with a bright, sunny day outside. The person becomes silhouetted against the sky. We can't make out his features. Those arc lamps correct the "balance" between the light on the actor's face and the bright sky. If we didn't use them, his face would go completely black[1].

Artificial lighting is a very time consuming and expensive process. In any movie production, regardless of scale, many lamps have to be transported and set up, requiring considerable human and material resources (movie crew,

[1]Although this explanation referred to cinema shot in film, it is the more pertinent for digital cinema because digital cameras are even more limited in this regard.

FIGURE 11.2: Artificial lighting is used in cinema and TV because of camera limitations: people present at the scene do not need these extra lights in order to see properly. From left ro right and top to bottom: images from [52, 50, 43, 51].

trucks, generators, etc.) and making for very complicated logistics: trucks and generators require lots of space, generators must usually be placed far off so that the noise they make does not compromise sound recording in the movie, and if the location covers a wide area or is difficult to access then the process of lighting becomes really hard and must be planned carefully. In general a very important limitation is that the director must decide prior to the shooting which objects are *seen*, as opposed to appearing completely dark or completely over-exposed, and at what level of detail. For example, in shooting *interior-day* scenes the director is usually forced to choose between showing what is inside or what lies outside the windows. The cinematographer then has to take time lighting the scene so as to follow the director's vision, and any changes at the post-production stage are difficult and expensive, mostly falling in the realm of special effects, e.g. if an object was shot too dark or too bright then its details are completely lost and it must be synthesized by a computer graphics artist, and of course re-shooting a scene to solve lighting problems is usually prohibitive.

11.2 High dynamic range imaging

In still photography, the problem of recovering the high dynamic range (HDR) radiances of a *static* scene has been solved since the late 1990's. There are several successful approaches, but probably the most popular and *de-facto* standard is that of Debevec and Malik [142], consisting of:

1. Taking a set of Low Dynamic Range (LDR) pictures of the scene with a fixed camera by varying the exposure time.

2. Using these LDR images to estimate the camera response function (for each color channel) by assuming that it is smooth and monotonic.

3. Computing the inverse of the camera response function and applying it to each LDR image, thus obtaining one radiance map estimate for each image.

4. Combining all radiance map estimates into a single HDR image by performing a pixel-wise weighted average, with larger weights given to radiance values that are far from the extrema values, where under- and over-exposure phenomena occur.

But if the scene is *dynamic*, with either camera or object motion, then the problem becomes quite challenging and remains open, with several approaches advocating for motion compensation and others proposing to operate directly on the combination or fusion step, e.g. see Ferradans et al. [165] and Hu et al. [200] and references therein.

For the generation of HDR video there are several possible approaches, as explained by Myszkowski et al. in their excellent book [290]:

- There are some user-made camera modifications where the camera is made to record an image sequence varying the exposure for each frame and then motion compensation is used (after shooting) to find correspondences among pixels [221] so that a traditional HDR-from-bracketing method such as the abovementioned [142] can be applied. There are two main issues with this approach. Firstly, it requires a very special type of camera because practically all video cameras (including professional cinema models) don't allow us to change the exposure time on a frame-by-frame basis. Secondly, HDR video created thus presents different kinds of artifacts, due to low frame-rate acquisition (frame interpolation is still an open problem in the context of film production; see Chapter 9), rapid motion, occlusions, very high dynamic range, etc. These visual artifacts must be corrected manually, which may explain the lack of popularity of these types of techniques.

- Obtaining different exposures in each single frame, by placing a mask of spatially-varying density before the sensor so that different pixels have different exposure values. The combination or fusion stage is akin to the demosaicking process seen in Chapter 6, and therefore the resulting HDR video might suffer from aliasing and other artifacts.

- Using two cameras with a beam-splitter rig. The cameras are identical, synchronized, aligned and have all the same parameter values except for the exposure time. The beam-splitter is a semi-transparent mirror that lets part of the light through to one camera and reflects the rest towards the other camera. Since the cameras are synchronized and the images perfectly registered, there are no issues with motion here and each pair of frames can be combined as if they were two differently exposed still images. The limitations of this approach come from the beam splitter, which may introduce color differences and which requires sensors with higher sensitivities.

- Solid state sensors, where the exposure time varies spatially or that directly compute the logarithm of the irradiance. Although some cameras capable of shooting HDR video with modified sensors do exist, they are built for specific applications (laboratory imaging, surveillance) or they are prototype, very expensive hardware [4].

Recalling the the f-stop terminology of Chapter 2, most professional and semi-professional DSLR cameras record images with a dynamic range of 12 f-stops, current digital cinema cameras of the RED family and the Black Magick camera support 13 f-stops, and the ARRI Alexa family reports up to 14 stops. These are large dynamic ranges, enough for many contexts, although not truly "high." For digital cinema, the only HDR imaging solution currently in use is the one developed by RED for some of their models [63]. It follows the first approach mentioned above: each frame is actually captured twice, in rapid succession, varying the exposure time, and then each pair is motion-compensated and combined. The main shortcomings of this technique are its reduced framerate and inability to handle large motions and occlusions.

11.3 Tone mapping

Displays are also limited in terms of dynamic range, so in order to handle HDR video on a common screen we need to perform a process called Tone Mapping (TM), which reduces the dynamic range of the input. The objective is usually to emulate as much as possible the contrast and color sensation of the real-world scene, i.e. achieve an image that looks natural, as opposed to trying to maximize the visible details that may make the resulting image appear artificial. This is the TM philosophy proposed by Ward et al.[375]:

"We consider the following two criteria most important for reliable TM: 1) Visibility is reproduced. You can see an object in the real scene if and only if you can see it in the display. Objects are not obscured in the under- or over-exposed regions, and features are not lost in the middle 2) Viewing the image produces a subjective experience that corresponds with viewing the real scene."

Figure 11.3 shows frames from different HDR videos tone-mapped for a regular display. We can see that the TM process has caused very noticeable color artifacts and halos around edges of moving regions (this same problem affects most HDR still images available on sites like Flickr). So we can see that, despite the intention being to obtain natural-looking pictures, the final results don't usually look realistic or natural due to limitations of the TM operators used.

FIGURE 11.3: Tone-mapped frames from HDR videos. From left to right and top to bottom: images from [40, 46, 55, 55, 49, 42].

Tone mapping is a very challenging open problem; see the book by Reinhard et al. [327] for an excellent survey. The literature on TM is abundant and many tone mapping operators (TMO's) have been proposed [164]: some based on perception and vision models [363, 122, 374, 337, 304, 325, 375, 76, 326, 355], some working directly on the contrast or the gradient field [364, 148, 161, 263], some allowing for user interaction [302, 253].

The best-performing methods are perceptually based and Ferradans et al. have very recently authored an algorithm [164] that considerably outperforms the state of the art in terms of contrast reproduction. The proposed method consists of two stages, the first one dealing with visual adaptation, the second one with local contrast enhancement. This design is based on psychophysical studies noting that while visual adaptation appears to be a completely retinal process, local contrast enhancement (and color constancy) must also involve the visual cortex, suggesting that the mechanisms involved in both processes are quite different.

For the visual adaptation stage, the authors make the observation that some of the best TM algorithms are based on the Naka-Rushton (NR) equation, which models the electrical responses of retina photoreceptors to flashed stimuli. Being a model of physiological responses, the NR equation *does not* model perception. On the other hand, the Weber-Fechner (WF) law *does* model the perception of detectable-difference thresholds for steady stimuli, but if used for TM the results are usually quite poor and definitely worse than those obtained with the NR equation. How can this be? The authors argue that the main reason resides in the fact that the NR curve saturates, as cones normally do, while the WF curve doesn't, it increases indefinitely, which would correspond to the case where cones never saturate. But this is for the WF curve in the case of steady stimuli: if the stimuli changes, for instance when we change our gaze, then the WF curve does indeed saturate. Therefore, the authors propose to use the saturating WF curve for TM: this curve has the nice properties of NR but, unlike NR, it is based on a perceptual model.

For the second step, of local contrast enhancement, the authors adopt the perceptual contrast enhancement method of Bertalmío, Caselles, Provenzi and Rizzi [95] and apply it to the output of the first stage (with a slight modification: the actual average of the first stage output is used for the dispersion term, instead of $\frac{1}{2}$ as in [95]).

The authors compare their method with the state of the art in TM using the metric of Aydin et al. [81]. This metric uses pyschophysical data to estimate the differences in luminance perception that an observer would have watching an HDR image and its tone-mapped version on a certain display with given viewing conditions. The differences, or "contrast distortions," are quantified and can also be visualized in a color-coded image, where gray represents no error, green means loss of contrast, blue means amplification of contrast, and red means contrast inversion. Figure 11.4 shows some examples. Quantitatively, this method produces a significantly smaller error (around 30% less) than state of the art algorithms, as well as a smaller number of pixels with distortion. As limitations of this method we can mention that it still is far from reaching zero contrast distortion, and its color behavior has not been quantified.

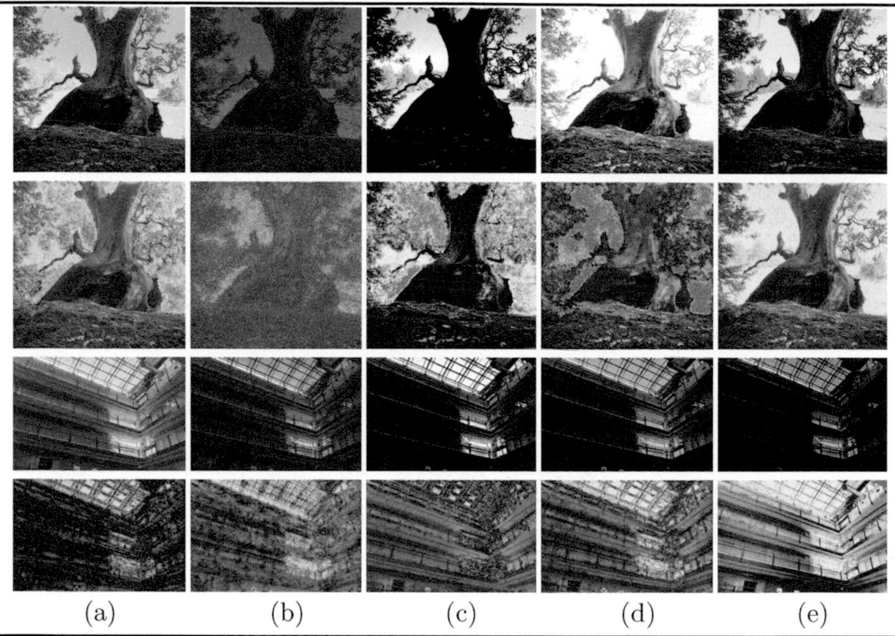

FIGURE 11.4: TM results (rows 1 and 3) and distortion maps (rows 2 and 4) for these TMOs: (a) Drago et al. [146], (b) Mantiuk et al. [263], (c) Pattanaik et al. [304], (d) Reinhard and Devlin [325], (e) Ferradans et al. [164]. Figure from [164].

11.4 Optimization of TM operators

In this section we present in some detail a recent algorithm by Cyriac, Batard and Bertalmío [136] that performs optimization of TM operators, reproducing excerpts and figures from that paper.

11.4.1 Minimization of the distance between images

11.4.1.1 Image quality metrics as non-local operators

Many tasks in image processing and computer vision require a validation by comparing the result with the original data, e.g. optical flow estimation, image denoising, tone mapping. Whereas measures based on pixel-wise comparisons (e.g. MSE, SNR, PSNR) are suitable to evaluate image denoising and optical flow estimation algorithms, they are not adapted to evaluate tone mapping. Tone mapping evaluation requires a measure that compares color sensation and detail visibility (contrast), which are not pixel-wise nor local concepts. This leads us to the following definitions.

Definition 1 (metric)**.** *Let $L, H\colon \Omega \longrightarrow \mathbb{R}$ be two images. We call* **metric** *an operator* **met** *such that $met(L, H)$ is of the form $met(L, H)\colon \Omega \longrightarrow \mathbb{R}^n, n \geq 1$.*

We say that a metric is **non local** *if the quantities*

$$\frac{\partial met(L, H)(x)}{\partial L(y)}, \frac{\partial met(L, H)(x)}{\partial H(y)} \tag{11.1}$$

do not vanish for some points $y \neq x$.

The two terms in (11.1) are infinite dimensional extensions of the notion of derivation with respect to the n-th component. For instance, the first term means that we compute the variations of the functional $L \longmapsto met(L, H)(x)$ with respect to the variations of the function L at the point y.

In this context, we can classify image quality measures into three categories. The first one is the class of metrics that are not non local, like MSE, PSNR and SNR measures. Indeed, the term (11.1) vanishes since such metrics are based on pixel-wise differences. The second category gathers the metrics that are non local and compares images of the same dynamic range (see e.g. [138] for LDR images, [262],[263] for HDR images, and [372] for images of any dynamic range). The last category contains the metrics that are non local and compares images of different dynamic range (see e.g. [81],[348]). In particular, the metric DRIM [81] belongs to both second and third categories since it is independent of the dynamic range of the images it compares.

Definition 2 (distance between images)**.** *A* ***distance*** *associated to the metric met is an energy E of the form*

$$E\colon (L, H) \longmapsto \frac{1}{|\Omega|} \int_\Omega \Phi(met(L, H)(x))\, dx \tag{11.2}$$

where $\Phi\colon \mathbb{R}^n \longrightarrow \mathbb{R}^+$ is a differentiable map.

11.4.1.2 Minimization of the distance: continuous formulation

From now on, we assume that H is fixed and met is a non local metric. We aim at finding the images L that minimize the distance with H. A necessary condition for L to be a minimum of the energy (11.2) is that L is a critical point of (11.2), i.e. the differential of the energy at this point in every direction is 0.

Proposition 2. *The images L that are critical points of the energy (11.2) satisfy*

$$\int_\Omega d\Phi \left(met(L, H)(x); \frac{\delta met(L, H)(x)}{\delta L(y)} \right) dx = 0 \qquad \forall y \in \Omega$$

The term

$$y: \longmapsto \frac{1}{|\Omega|} \int_{\Omega} d\Phi \left(met(L,H)(x); \frac{\delta met(L,H)(x)}{\delta L(y)} \right) dx \tag{11.3}$$

is the gradient $\nabla(E)$ of the functional $L: \longmapsto E(L,H)$.

It is not trivial to obtain mathematical properties of functionals (11.2) because of the nature of the operators *met*. Hence, it is hard to establish accurate numerical schemes in order to get the minima of the energy (11.2). For this reason, we deal here with the gradient descent algorithm, which has the property of being applicable to a large class of functionals. Given an image L_0, we consider the scheme

$$L_{k+1} = L_k - \alpha_k \nabla(E)(L_k), \quad L_{|t=0} = L_0. \tag{11.4}$$

However, such a method may reach critical points that are local minima and not global minima. Hence, the initial condition has to be carefully chosen.

11.4.1.3 Minimization of the distance: discrete formulation

We assume that L and H are of size $M \times N$. We define the (discrete) distance between L and H as

$$E(L,H) = \frac{1}{MN} \sum_{m=1}^{M} \sum_{n=1}^{N} \Phi(met(L,H)(m,n)). \tag{11.5}$$

In this section, we construct a discrete counterpart of the gradient (11.3). The main task is to define a discrete counterpart of the term (11.1). In particular, the term $\delta L(y)$ yields the construction of pixel-wise intensity increments.

11.4.1.4 Pixel-wise intensity increments

There is not a unique choice to compute pixel-wise intensity increments. The trivial approach consists of considering the value $L_+(i,j): = L(i,j) + k$ as forward increment and $L_-(i,j): = L(i,j) - k$ as backward increment at a pixel (i,j), for a fixed $k \in \mathbb{R}^{+*}$. However, we claim that such an approach is not adapted to the problem addressed here. Indeed, because the metric *met* in (11.1) involves perceptual concepts, we intuit that the increments should involve perceptual concepts too.

In [342], a perceptual distance between pixels that takes into account the Weber law principle has been constructed. In particular, the authors construct a "perceptual gradient"

$$\nabla_W I: = \frac{\nabla I}{I} \tag{11.6}$$

for image denoising purposes.

Then, we define the backward $L_-(i,j)$ and forward $L_+(i,j)$ increment at the pixel (i,j) by requiring their perceptual distance (in the sense of Shen) with $L(i,j)$ to be equal, i.e.

$$\frac{L_+(i,j) - L(i,j)}{L(i,j)} = \frac{L(i,j) - L_-(i,j)}{L(i,j)} = k$$

for some $k > 0$.

11.4.1.5 Expression of the discrete gradient

Given a constant $\lambda \in]0,1[$, we construct for each pixel (i,j) two images $L_+(i,j)$ and $L_-(i,j)$ defined as follows

$$L_+(i,j)\colon (m,n) \longmapsto L(m,n)\Big(1 + \min(1/L(i,j) - 1, \lambda)\,\delta((m,n) - (i,j))\Big)$$

$$L_-(i,j)\colon (m,n) \longmapsto L(m,n)\Big(1 - \min(1/L(i,j) - 1, \lambda)\,\delta((m,n) - (i,j))\Big)$$

where δ is the Dirac delta function.

Remark 3. *By the use of the increment* $k = \min(1/L(i,j) - 1, \lambda)$, *we impose that the values of the images* $L_+(i,j)$ *do not exceed 1. Moreover, the assumption* $\lambda \in]0,1[$ *imposes that the values of the images* $L_-(i,j)$ *do not become negative.*

Then, we define the discrete version of the term (11.1) as

$$\frac{\delta met(L,H)(m,n)}{\delta L(i,j)} := \frac{met(L_+(i,j),H)(m,n) - met(L_-(i,j),H)(m,n)}{2min(1/L(i,j) - 1,\, \lambda)} \tag{11.7}$$

from which we derive each component (i,j) of the discrete gradient:

$$\frac{1}{MN}\sum_{m=1}^{M}\sum_{n=1}^{N} d\Phi\left(met(L,H)(m,n); \frac{[met(L_+(i,j),H) - met(L_-(i,j),H)](m,n)}{2min(1/L(i,j) - 1,\, \lambda)}\right). \tag{11.8}$$

Finally, we perform the gradient descent (11.4) where $\nabla(E)(L_k)$ is given by (11.8).

11.4.2 Metric minimization for TMO optimization

In this section, we make use of the minimization approach introduced above in order to improve tone mapping operators relative to a given metric. This is a general approach that can be applied to any TMO. Whereas any dynamic range independent metric that can be defined as a non-local metric in the sense of def.1 can be used, we focus in our experiments on the metric DRIM [81].

11.4.2.1 Dynamic range independent metrics (DRIM)

The metric DRIM compares images of any dynamic range. For applications to tone mapping evaluation, the inputs are an HDR reference image H and its LDR tone-mapped image L. The purpose of the metric is to consider the perception that a viewer would have of both scenes relying on psychophysical data, and to estimate at each pixel the probabilities that distortions appear. The distortions are based on the detection and classification of visible changes in the image structure. Three types of distortion are considered: loss of visible contrast, LVC (contrast visible in the HDR image and not in the LDR image), amplification of invisible contrast, AIC (details that appear in LDR image that were not in the HDR image) and contrast reversal, and INV (contrast visible both in the HDR and LDR images, but with different polarity).

Each type of distortion can be represented as a $[0,1]$-valued function on the image domain $\Omega \subset \mathbb{R}^2$. Then, we define a metric met (in the sense of Def. 1) between HDR and tone mapped LDR images by

$$met\,(L,H) : = (LVC, AIC, INV) \tag{11.9}$$

from which we define the distance $E(L,H)$ between L and H associated to met as

$$E\,(L,H) = \frac{1}{MN}\sum_{m=1}^{M}\sum_{n=1}^{N}\sqrt{LVC\,(m,n)^2 + AIC\,(m,n)^2 + INV\,(m,n)^2}. \tag{11.10}$$

Note that the function ϕ of Def. 2 we take in the expression of (11.10) is the Euclidean norm on $\mathbb{R}^3$. Applying formula (11.8) to the energy (11.10), we obtain the expression of the gradient of the functional $L \longmapsto E(L,H)$

$$\nabla\,(E\,(L,H))\,(i,j) = \frac{1}{MN}\sum_{m=1}^{M}\sum_{n=1}^{N}\frac{\langle met\,(L,H\,(m,n))\,, \frac{\delta met(L,H)}{\delta L(i,j)}\,(m,n)\rangle}{\phi\,(met\,(L,H)\,(m,n))} \tag{11.11}$$

where the term $\frac{\delta met(L,H)}{\delta L(i,j)}$ is given by (11.7). Then we perform the gradient descent method (11.4) in order to reduce the distance (11.10) between the tone mapped image L and the HDR source H.

The existence of global extrema of the energy (11.10) is ensured. Indeed, the energy is bounded below by 0 and above by $\sqrt{3}$ since the metric met is bounded below by 0 and above by $\sqrt{3}$. Moreover, the set $L^2(\Omega;[0,1])$ on which we minimize the energy is closed and bounded.

However, local extrema might exist because we do not have information about the convexity of the energy. Hence, the gradient descent method we apply may converge to a local minimum that is not global.

Remark 4. *Even if the theoretical lower bound of the energy is 0, it is barely conceivable that this bound might be reached. Indeed, experiments show that we have $E(L,L) > 0$ and $E(H,H) > 0$ when we test the metric DRIM on any*

FIGURE 11.5: The preprocessing stage. From left to right: input TM image (distance =0.717), output image (distance=0.682), distortion map of the input, distortion map of the output. In this case, the preprocessing reduces the TM error by 5%. Figure from [136].

LDR image L or HDR image H of non constant intensity values. Hence, we claim that the situation $E(L, H) = 0$ does not occur with the DRIM metric. As a consequence, we do not know if the critical point our algorithm reaches is a local or global minimum.

11.4.2.2 Preprocessing

To increase the chance that the algorithm converges to an image that has the smallest distance to H, i.e. a global minimum, we apply a preprocessing on the original tone mapped image L in order to get an initial condition of our algorithm that is closer (in terms of distance) to H than L. The method we propose relies on the intuition that high values of LVC might be reduced by application of local sharpening whereas high values of AIC might be reduced by local Gaussian blurring. Hence we perform local Gaussian blurring and unsharp masking [68] to the initial LDR image depending on the values of the function $LVC - AIC$.

Denoting by L_{smooth} a blurred version of L and defining L_{sharp} as

$$L_{sharp} = L + \alpha (L - L_{smooth}) \tag{11.12}$$

for some constant α, we define the image L_{new} as

$$L_{new}(i,j) = \begin{cases} L_{sharp}(i,j) & \text{if} \quad LVC(i,j) - AIC(i,j) > 0 \\ \\ (1-\beta)L(i,j) + \beta L_{smooth}(i,j) & \text{if} \quad LVC(i,j) - AIC(i,j) < 0 \end{cases} \tag{11.13}$$

Experiments (see Table 11.1) show that we do have $E(L_{new}, H) < E(L, H)$.

11.4.2.3 Experiments

As the metric DRIM only takes into account the luminance information of the images, we first convert the LDR color image into a luminance map. It is

FIGURE 11.6: The final output. First column: input TM images (from top to bottom: Drago et al. [146], Reinhard et al. [325], Mantiuk et al. [264], Ferradans et al. [164]). Second column: final output images. Third column: distortion maps of input TM images. Fourth column: distortion maps of final outputs. See Table 11.1 (image "Harbor") for the corresponding distances. Figure from [136].

performed by mapping its RGB values into the Lab color space and extracting the L-channel. Then inverse gamma correction is applied.

We perform the preprocessing described above, with the following parameters: the parameter σ of the Gaussian smoothing kernel is set to 0.62, and the constants α, β are respectively 0.7 and 0.5. These values provide good results and have been fixed for all the experiments in this paper.

Then, we apply the gradient descent algorithm of Section 11.4.1.5, where the initial condition is the output of the preprocessing. Because the variational

FIGURE 11.7: Comparison between the preprocessing stage and the final output. Top left: input TM image (distance = 0.708). Top middle: output of preprocessing stage (distance = 0.615). Top right: final output (distance = 0.533). Bottom row: corresponding distortion maps. For this image our method reduces the TM error by 25%. Figure from [136].

problem we propose is non-local, the gradient descent algorithm is very time-consuming. In order to decrease the execution time of the algorithm, rather than computing the sum in Equation (11.8) over the whole image, we only consider 50×50 neighborhoods. The algorithm stops when a local minimum is reached, i.e. when the energy does not decrease anymore. The output LDR color image is obtained by adding the components a,b of the initial TM image to the output LDR luminance map.

The evaluation of our algorithm is two-fold: global and pixel-wise. As a global measure, we consider the total energy (11.5) of the output (see Table 11.1). As a pixel-wise measure, we make use of the distortion map in [81]. The basic idea is to encode contrast distortions by colors: green hue represents LVC, blue hue stands for AIC and red hue indicates INV, the saturation encoding the magnitude of the corresponding distortion, whereas the intensity corresponds to the intensity of the HDR source image (up to rounding). At each pixel, the maximum of the three distortions is computed, and the corresponding color is displayed. If the maximum is lower than 0.5, then the saturation is set to 0.

The first experiment consists of evaluating the preprocessing. In Figure 11.5 we show an output color image of the preprocessing, as well as its distortion map. We have applied the Formula (11.13) to a TM image obtained with the method of Ferradans et al. [164]. The HDR source image is taken from the MPI database [62]. We observe that the LVC distortion has been reduced, whereas the INV distortion has increased a bit. As we can see in formula (11.13), the preprocessing is only devoted to reduce the LVC and AIC distortions, and does not take into account the INV distortion. This might explain why the INV distortion tends to increase. In Table 11.1, we present results of the preprocessing tested on images of the Fairchild database [66] for different TMOs, i.e. Ferradans et al. [164], Drago et al. [146], Reinhard et al. [325], Mantiuk et al. [264]. The images have been rescaled to 200×200 pixels in order to speed the algorithm up. Average results have been computed over 5 images of the dataset. The results confirm that the preprocessing reduces the distance with the HDR source image.

In the second experiment, we evaluate the final output of our method described above. In Figure 11.6 we show some results of our algorithm tested on the different TMOs mentioned above. The HDR source is the image "Harbor" taken from the Fairchild database. The distortion maps show that the algorithm drastically reduces the distortion LVC, but the INV tends to remain. Table 11.1 shows the distances of the initial TM images and output images with the HDR source image, as well as for other images of the Fairchild dataset. As expected, the final distance is lower than the initial distance. Another interesting point is the observation in Figure 11.6 that the contrast in the output color images is higher than the one in the initial TM images.

At last, we compare the output of the preprocessing stage with the final output of our method. In Figure 11.7, we compare distortion maps. The HDR source is the image "Dam" from the Fairchild database. The input TM image is provided by the TMO of Reinhard et al. [325]. The corresponding distortion map shows a high loss of contrast. We observe that the preprocessing stage reduces it to a great extent, and the gradient descent algorithm applied to the output of the preprocessing reduces it more. However, we see that neither the preprocessing nor the gradient descent algorithm achieves the reduction of the INV distortion. The results on Table 11.1 confirm that applying the gradient descent algorithm to the output of the preprocessing provides better results than applying only the preprocessing.

11.5 Final remarks

Recording movies in HDR and performing TM of the resulting footage are two challenging open problems.

All HDR video creation techniques present issues, most yielding visible artifacts in common conditions such as with rapid motion, occlusions, very high dynamic range, etc.

TABLE 11.1: Distance at each stage of our method computed with DRIM [81]. Notation: I (initial), P (Preprocessing), F (final), Imp (improvement). Table from [136].

Image	TMO	I	P	F	Imp(%)
Harbor	Drago et al.	0.681	0.608	0.502	26.25
	Reinhard et al.	0.676	0.607	0.493	27.05
	Mantiuk et al.	0.528	0.487	0.424	19.75
	Ferradans et al.	0.560	0.509	0.438	21.46
Dam	Drago et al.	0.767	0.682	0.584	23.90
	Reinhard et al.	0.708	0.615	0.533	24.70
	Mantiuk et al.	0.726	0.649	0.564	22.25
	Ferradans et al.	0.691	0.630	0.542	21.50
Average	Drago et al.	0.672	0.606	0.525	21.91
	Reinhard et al.	0.636	0.564	0.464	27.09
	Mantiuk et al.	0.619	0.564	0.493	20.40
	Ferradans et al.	0.609	0.566	0.491	19.39

With TM operators the situation is better in the sense that state-of-the-art TMOs produce results without noticeable artifacts; the downside is that TMOs are hard to evaluate and compare, because all the available metrics (such as [262, 237, 373, 348, 81]) have shortcomings: most are designed for still images, some of these metrics are suitable only for comparing LDR images, others do not model contrast masking but only changes in visibility, none takes into account the saturation in the perception of high luminances under non-steady stimuli, the experimental validation is usually lacking, and none of these works deals with color distortion. In the literature there does not exist any perceptually-based, quantifiable metric for *color* distortion in tone mapping (the metrics above measure *contrast* distortion), probably due to the difficulty of combining perceptual data with the particular color-space characteristics of any given color display, so this remains a fully open problem.

Chapter 12

Stereoscopic 3D cinema

12.1 Introduction

In the past few years there has been a resurgence of stereoscopic 3D (S3D) cinema, with many voices in the industry arguing that this time 3D is here to stay. The truth is that stereography, as created by Wheatstone in 1838, predates photography, and 3D movies were invented by William Friese-Green in 1889, some years before cinema became popular [408]. S3D cinema for a general audience started in 1922, faded out quickly, lived a golden age between 1952 and 1954, disappeared again, had a short revival in the 1980s, went away, and is now enjoying a mainstream comeback that started in 2003 [7] but had its real major boost with James Cameron's "Avatar" in 2009.

The principle of S3D cinema consists of presenting each of our eyes with a different image, thus triggering in our brains all the visual mechanisms that allow us to estimate the distance of objects and therefore to perceive the world as a three-dimensional space, rather than a flat representation. The method by which a single image (view) is presented to each eye has changed over the years.

S3D movies in the 1920s were *anaglyph*: each film frame combines the two views, which are color-filtered (e.g. the left view with red, the right view with cyan) and superimposed; the viewer must wear anaglyph glasses, where for each eye there is a translucid plastic strip with the same color as the corresponding view so that the image for the other eye is (mostly) filtered out. Anaglyph is an extremely simple and cheap technology, and it has also made a comeback, not in cinemas but for watching Internet videos and pictures in 3D, and also for printed books.

S3D movies from the mid-1950s until the present day use polarization to create the stereoscopic effect. Polarization is the orientation of the plane on which the light is traveling. Some processes originate unpolarized light (a mixture of all polarizations), others generate polarized light. We are not sensitive to polarization of the light, only to its wavelength and intensity, but there are materials with which polarizing filters can be manufactured: these filters block all light of one polarization, let all light pass that has a polarization perpendicular to that of the light that is being blocked, and pass some light with in-between polarizations. Polarizing filters are very useful in photography because they intensify the colors of a scene, make white clouds stand out against the sky, and reduce glare and reflections from water or glass: reflected light (except from metal) is polarized, so by rotating a polarizing filter the photographer can cancel a reflection and see what's beyond the reflecting surface. In S3D cinema projection the two views have different polarizations, and the viewer wears polarized glasses consisting of different polarizing filters, each letting through the light from one image while blocking the light from the other image.

Present day 3DTV displays are of three different kinds. Active 3D displays quickly alternate left and right views, so instead of 50 or 60 pictures per second as in a regular TV they will display 100 or 120 pictures per second. They require the use of active glasses, which activate independent shutters for each eye and which must be synchronized with the 3DTV so that when the TV is displaying the left view then the shutter for the right eye is closed, and viceversa. These glasses are quite expensive and require batteries. Passive 3D displays are based on polarization: one view is presented on the even lines, the other on the odd lines, and the display has a filter that polarizes each line; the system requires the use of polarized glasses, which are cheap [25]. A third kind is that of autostereoscopic displays, which don't require the use of glasses because they use a parallax barrier to present each eye with only one image.

While many people have the impression that 3D movies in the 1950s were just exploitation films in anaglyph and this is why they went out of fashion, the truth is that 3D films of the 1950s used polarization just as they use it today, and even Alfred Hitchcock directed a 3D movie, "Dial M for Murder" [408]. The main problems of S3D cinema seemed to be related to visual discomfort, produced by inefficient image separation techniques, excessive disparities among views, which made binocular fusion impossible and caused double vision, and projector misalignment [241] and dis-synchronization [7]. All of these issues are carefully treated nowadays, but still the comfort levels in S3D cinema have not reached those of regular, 2D cinema, and some critics argue that this will never happen, that S3D will disappear again this time [3, 35] because of the problems inherent to stereoscopic cinema. These problems are mainly the accomodation/vergence mismatch (discussed below) and the extra effort required from the viewer, who is forced when watching a stereoscopic movie to constantly and passively adapt her eyes as commanded from the screen, whereas real-life 3D is not a passive but an active experience [35]. In any case the interest in S3D cinema appears to be lightly winding down, as revenue and number of 3D movie releases decrease [58], sales of 3DTVs grow very slowly [38], and even the BBC has abandoned 3D transmissions [65]. There are several other factors as well, like the higher costs for 3D cinema tickets and 3DTV displays, the requirement of wearing glasses, which may be uncomfortable for individuals already wearing prescription glasses, the reduced luminance of the images in theaters, and in general the rather large cost/reward ratio of the S3D cinema experience.

12.2 Depth perception

Our visual system allows us to estimate the distances of objects (i.e. their "depth") by combining many different visual cues, some arising independently in each eye (monocular depth cues), some given by relating information coming from both eyes (binocular depth cues).

12.2.1 Monocular depth cues

Monocular cues are very powerful; we are very proficient in estimating depth information of a scene from just looking at a (flat) picture of it. Being monocular, the information they provide is the same if we close one eye or if we look at a scene with both eyes. Among these cues we may cite [14]:

- *Motion parallax.* When we move our head, faraway objects move less in our retinal image, while closer objects move more.

- *Depth from motion.* A moving object that becomes larger is perceived as approaching us, reducing its depth; conversely, if an object moves and becomes smaller it is perceived as receding, increasing its depth.

- *Perspective.* When in the scene there are lines that we know to be parallel, they provide us with a depth cue because they get progressively closer with increasing distance.

- *Size gradient.* Image size depends on object size and object distance, so if two objects of the same (real) size appear to us as having different sizes, the larger one must be closer. The same holds for a single object with a textured pattern: the regions where the pattern appears larger are closer to us (e.g. the case of a brickwall in perspective).

- *Familiar size.* If we are familiar with the actual size of an object, we can combine this knowledge with the image size of the object to estimate its absolute depth.

- *Accommodation.* When we look at a far-away object, our eye can be in a relaxed state and the image of the object is formed on the retina. On the other hand, if the object is close and the eye is relaxed, its image will form behind the retina so we would be seeing it blurry; our ciliary muscles must contract the eye lens, making it thicker and hence reducing its focal length so that the image forms on the retina. The sensations arising from contracting and relaxing the ciliary muscles are processed in the visual cortex to assist the estimation of depth. See Figure 12.1.

- *Occlusion.* If we know an object and it appears partially covered by another, then we perceive the occluding object as being closer to us, and the obstructed object as being at a larger distance.

- *Lighting and shading.* Knowing the direction where the lighting is coming from (or assuming it comes from above), we can use shading information to estimate shapes and distances.
- *Defocus blur.* If and object lies outside our DOF (depth of field) it will appear blurry, out of focus. Therefore, it will be either much further or much closer than the object we're focusing on.
- *Texture or grain.* The closer we are to an object, the more detail we can see on its surface.

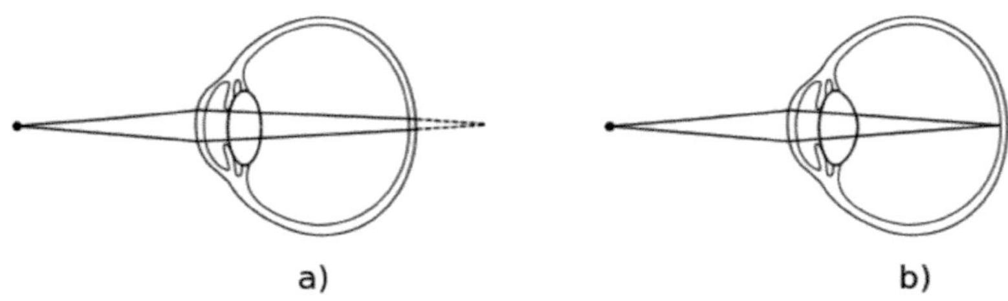

FIGURE 12.1: Accommodation. (a) If we look at an object that is close and the eye is relaxed, the image will form behind the retina and the object appears blurry. (b) Our ciliary muscles contract the eye lens, making it thicker and hence reducing its focal length so that the image forms on the retina.

12.2.2 Binocular depth cues

The two main binocular depth cues are *vergence* and *stereopsis*.

When we look at an object, our eyes move so that their lines of vision cross at the object: this movement is called *vergence*. If the object is close by the eyes move in opposite directions towards each other and the motion is called convergence; if the object is far away, the eyes move away from each other and the motion is called divergence. Vergence requires stretching or relaxing the extraocular muscles, producing sensations that are used by our brain as a depth cue [14]. In normal conditions, vergence is intrinsically and reflexively linked to accomodation [241]: when we look at an object, our eyes move to converge on it and at the same time they accomodate so that the object is in focus.

Stereopsis is the depth cue given by the disparity between corresponding points in the two retinal images. Since our eyes are some 6 cm apart our retinal images are slightly different, and the same scene point will produce an image at a different location on each retina. The distance between these two image points will be small for scene points that are far away, and large for scene points that are close to our eyes. Therefore, our visual system fuses both retinal images into one and from the disparities among corresponding points we are able to estimate their depth. This is the fundamental principle of stereoscopic photography and 3D cinema. When we fixate on an object, its images will fall on corresponding parts of the retinas and therefore its retinal disparity will be zero [241]. Points in space with zero retinal disparity lie on an arc around the viewer called the *horopter*. Points before the horopter will have crossed disparity, appearing at the right on the left image and at the left on the right image; points behind the horopter will have uncrossed disparity. See Figure 12.2. Under binocular fusion, and for a given vergence, the accomodation has a certain DOF range where it can vary and still objects are perceived properly. The same happens for a given accommodation and varying vergence. The two retinal images can be fused into a single image for all points within a small region around the horopter, called Panum's fusional area, whose limits go from 0.1 degrees of maximum disparity at the fovea (the retina's central spot) to $\frac{2}{3}$ of a degree at 12° from the fovea [241].

12.2.3 Integrating depth cues

In everyday conditions, the brain combines depth information from multiple depth cues in order to reduce ambiguity and achieve a consistent depth perception; but when watching stereoscopic images, i.e. when a different planar image is presented to each eye, then the depth cues may conflict and studies show that occlusion is dominant over all other cues and is only approached by binocular disparity, while accommodation and vergence have only a small effect [241].

Individual differences and gender and age differences affect the preference for stereoscopic images [241]:

- People with smaller inter-pupillary distance perceive more depth and reach disparity limits (above which binocular fusion can't be performed) faster.

- AC/A is defined as change in convergence due to accommodation per change in accommodation. People with high AC/A have trouble with binocular fusion.

- A larger pupil diameter decreases the quality of the final stereoscopic image due to an increase of spherical aberrations and a reduction of the DOF.

- The ability to perform accommodation starts to decay at the age of 40 and it's virtually gone after the age of 55.

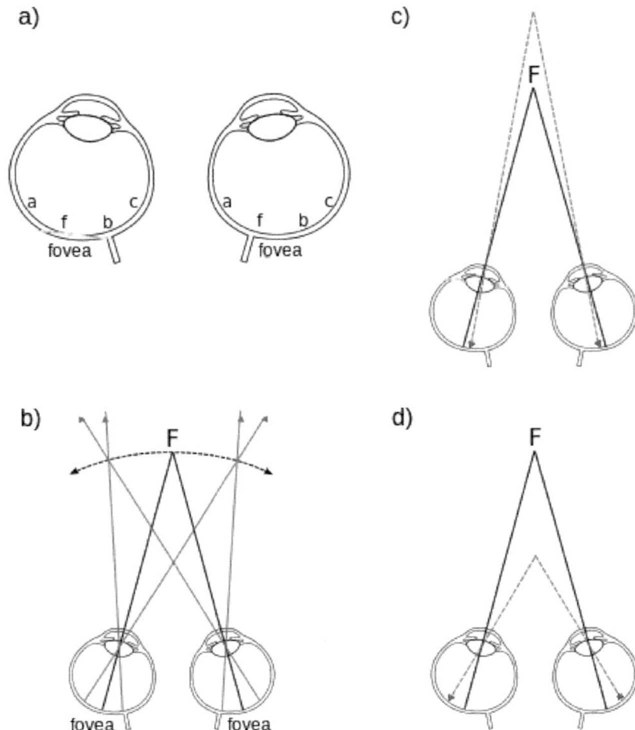

FIGURE 12.2: Retinal disparity. (a) The fovea and three pairs of points at corresponding retinal locations. (b) Point F lies on the horopter (dashed curve). The images of points located farther (c) or closer (d) than the horopter do not fall at corresponding retinal locations. Adapted from figure 6.17 of Brian Wandell's "Foundations of vision" [1].

12.3 Making S3D cinema

There are three ways to produce an S3D movie. In increasing order of cost:

- in 3D animation, rendering each frame twice, from two slightly different points of view, thus generating the stereoscopic images;
- in live action, converting in post-production a conventional 2D picture into 3D by using the original movie as one view and synthesizing the other view with software tools and manual interaction;
- in live action, using two identical, synchronized cameras placed one next

to the other, each recording a view corresponding to the view of one of the eyes.

12.3.1 3D animation

In animation, "3D" means that what is shown on the screen are not planar drawings, but the result of rendering on the image tri-dimensional representations of objects; each object in 3D animation is a computer model with 3D coordinates that is placed in a 3D scene and then its picture is taken using a "virtual camera" located in the same 3D world. We can see that with very little extra effort two views can be generated instead of just a single view, so the cost of producing an animated movie for regular 2D projection or for stereoscopic projection is very similar.

12.3.2 2D-to-3D conversion

This task requires generating, for each frame of the movie, a slightly different new frame corresponding to the picture a second camera would have taken if it had been placed in the shoot, side to side with the other camera. Of course there isn't a formal solution to this problem: we can't recover the 3D geometry of a scene from a 2D image of it.

What is done in practice is that a stereographer puts "notes" on objects specifying their depth; then, an artist manually extends these notes and paints the depth map. In some cases (e.g. vehicles, people's heads), 3D models of the objects are built. This requires rotoscoping (segmentation plus tracking) of objects, which usually takes up to 80% of the budget for 2D-to-3D conversion. The process is assisted by software but requires heavy and skilled manual interaction, for instance an artist may carefully mark the contour of an actor in a keyframe and a software tool then tracks this contour as the actor moves frame after frame. With a view and its associated depth map, a new view can be generated, as we'll see in Section 12.7.

An important limitation is that of occlusions: in the new view that must be created, the slightly different perspective makes visible regions that were not visible in the original image. When we move an object in the new view to simulate the new perspective, what was behind that object becomes disoccluded and therefore it's an image gap that must be filled in: this is the inpainting problem. See Figure 12.3. Inpainting techniques provide very good results for still images, but there isn't yet an (automatic) inpainting method for image sequences that is capable of reaching the visual quality required in cinema, without noticeable temporal artifacts, as we discuss in Chapter 14.

Currently about half of the mainstream S3D movies are shot conventionally and post-converted to 3D. The main advantages of this approach are:

- It allows for an easier, much cheapier shoot. Native S3D shoots are 20% to 50% more expensive than a regular 2D shoot.

- The cost is predictable, depending on the intended final quality, and there's no risk of going over budget. A native 3D shoot, on the other hand, is subject to unpredictable increases in costs, as any movie shoot is.

- The final cost is still smaller than that of a native 3D shoot.

Because of these reasons, the trend would appear to be shifting towards 2D-to-3D conversion: most recent high-budget action movies released in 3D have been shot in 2D and post-converted [29].

As for the disadvantages, we can mention that it requires a lot of work, it may not yield good results in scenes with transparencies or motion blur, and then there's the issue of inpainting. Therefore, converting 2D to 3D is a very labor intensive endeavor in which the role of artists and technicians is fundamental. As an example, the 3D re-release of the film "Titanic" took two years of work by a team of 300 people, and cost 18 million dollars [37].

FIGURE 12.3: Two stereo views. If the left view were to be created using just the right view, all the areas marked in red should be inpainted.

12.3.3 Stereoscopic shooting

Live-action, professional 3D movie production is an even more involved, delicate and costly process, the best suited for achieving a *realistic* depth (though not necessarily *artistic*).

For starters, it is crucial that the twin cameras used in a stereo set-up perform in exactly the same way. They have to be the same model, produced in the same year (even preferably with consecutive serial numbers), running the same software, with all presets reset to factory defaults. Also, all camera parameters must be matched: white balance, sensitivity, shutter angle, exposure, focus distance, focal length, frame rate, and gain [270]. Optics must be paired as well, but due to manufacturing processes it is impossible for lens makers to produce two identical lenses and discrepancies must be solved with production and post-production tools [270]. The twin cameras are mounted on a rig, and there are two possible configurations: "side-by-side" with a parallel rig, and "mirror" with a beam-splitter rig. The latter is very popular since close-up shots require bringing the camera interocular down to a few millimeters, and the large cameras and optics of professional productions make parallel layouts inadequate [270]. In the beam-splitter rig the cameras are placed at 90 degrees and a semi-transparent mirror (the beam splitter) reflects the scene into one camera while allowing the other camera to see through the mirror. Apart from physical considerations (the mirror is fragile, dust prone, reduces light by one f-stop and requires the image shot off the mirror to be flipped) the beam splitter produces color differences among the views, which again can only be dealt with in post-production [322]. Due to these artifacts from the mirror (typically in the right eye when shooting with a beam-splitter rig) the left eye image is usually considered the master image, and is generally used when creating 2D from 3D material [322].

We could think that in order to achieve a realistic 3D effect we must ensure that the distance between the two cameras is exactly the average distance between our pupils (around 63 mm [241]), and that the focal length of the camera lens is exactly the average focal length of the eye (around 17 mm [22]). But this would be *very* limiting in artistic terms. In the language of cinema, different focal lengths are used to convey different emotions, to say things about the characters, to underline some elements of the story and help the narration; as we saw in earlier chapters, changing the focal length alters the relative distance of the objects in the scene (e.g. we can bring the main character closer to the background by using a longer lens), the angle of vision (we can see much more of a landscape with a shorter lens), the depth of field (increasing the focal length makes a larger part of the image become blurry). Therefore, since we are used to seeing movies where the focal length is neither 17 mm nor constant, we can't expect S3D movies to have this limitation. Also for artistic reasons the distance between cameras is not kept constant at 63 mm, because modifying this distance is what enhances the stereo effect, varying the total perceived depth of the scene. But changing interaxial separation (inter-camera distance) and focal length must be done very carefully and is the source of several challenging problems, as we'll briefly see in the following section.

12.4 Parallax and convergence

The parallax of an object is the screen distance between its left and right images; more precisely, it's the difference of the horizontal coordinate of its right image minus the horizontal coordinate of its left image. As we can see in Figure 12.4, the sign of the parallax is what makes the viewer perceive the object before, at or beyond the screen. With negative parallax, the left image of the object is at the right of its right image (i.e. if we look at this object our eyes will be crossed). When the two camera images of a subject are superimposed on top of each other and are aligned, the subject in question has zero parallax, and will appear to be at the same distance from the viewer as the screen onto which it is being projected [60]. With positive parallax, the left image is to the left of the right image and the object appears beyond the screen.

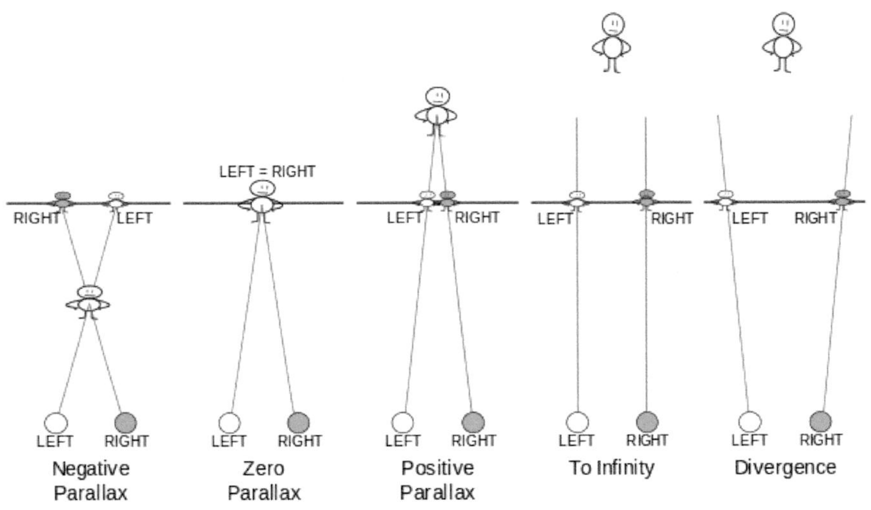

FIGURE 12.4: Parallax.

If the parallax is positive and equals the inter-ocular distance of 63 mm, then the lines of vision are parallel and the object is perceived as being at infinity; see Figure 12.5. This value of 63 mm is the maximum positive parallax (MPP), because a larger value would imply that in order to achieve binocular fusion on this object our eyes should look away from each other and this is a very discomforting experience that puts a lots of strain on the viewer. Therefore, the MPP value of 63 mm is a rule that should be respected [60]. Maximum negative parallax (MNP), on the other hand, does not impose such a hard limit because convergence (eye-crossing) is not so stressing as divergence; ideally, MNP is defined as equal to the MPP in which case the object is perceived as being halfway to the screen; see Figure 12.5.

In practice both MPP and MNP are expressed not in millimeters but in pixels, and therefore they are a function of the screen size: a smaller screen allows a larger MPP in pixels, and a larger screen allows a smaller MPP (because its pixels are larger). For instance, the same movie with a horizontal resolution of 1920 pixels will have an MPP of 200 pixels when shown on a TV monitor and of 13 pixels when shown on a large cinema screen [60], so the left and right view shouldn't be the same for a theatrical release and for Blu-ray, for instance.

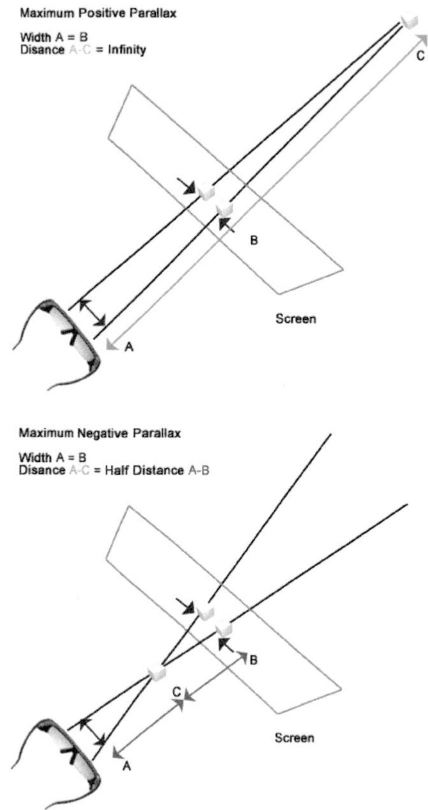

FIGURE 12.5: Maximum positive (left) and negative (right) parallax. Figures by Steve Shaw of Light Illusion [60].

Convergence is determined by the angle formed by the optical axes of the cameras, and the convergence point is where the left and right images are aligned. The convergence point appears to be on the screen, whereas objects before the convergence point appear in front of the screen and objects behind the convergence point appear beyond the screen [60]. This is *camera* convergence we're talking about: the viewer's eyes may converge and do converge elsewhere, onto virtual objects before or behind the screen, otherwise we wouldn't enjoy a full-range stereoscopic-depth experience. Modifying the convergence is equivalent to changing the distance between left and right image, also called horizontal image translation or HIT. This in turn creates a change in parallax, and results in moving the whole 3D scene forward or backward. It's only a global shift in depth, respecting the distances among scene objects. See Figure 12.6.

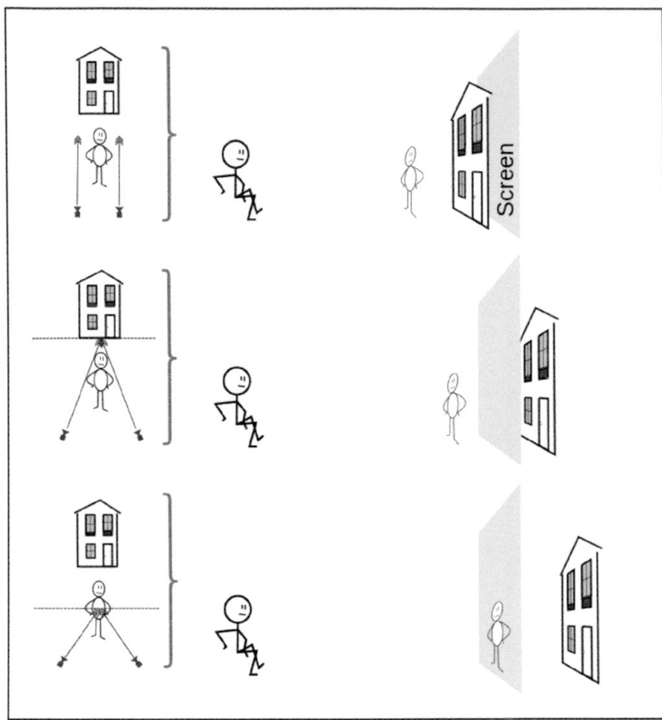

FIGURE 12.6: Modifying the convergence creates a depth shift. Figure adapted from [270].

Convergence can be achieved when shooting, by "toeing-in" the cameras so that their optical axes cross at the desired point ("converged shooting"), or afterwards, by shooting using parallel optical axes ("parallel shooting") and later in postproduction shifting the images horizontally with what is called "re-convergence HIT." The preferred option is the latter, parallel shooting followed by re-convergence HIT, because converged shooting may cause keystone artifacts (see Figure 12.7), and also because converged shooting requires changing the convergence value constantly in the same shot as the object of interest moves to-and-fro in the scene, and this may add a lot of time to a shoot [60].

With parallel shooting all objects in the scene have negative parallax, because the point of zero-parallax is at infinity. Therefore, all objects will appear to be *before* the screen. This creates a very important problem for objects near the boundary of the frame that are partially occluded: the monocular cue of occlusion says the object is behind the screen, while the binocular clue of stereopsis says the object is in front. This conflict of depth cues is to be avoided at all costs, because apart from the discomfort it provokes in the viewer, very

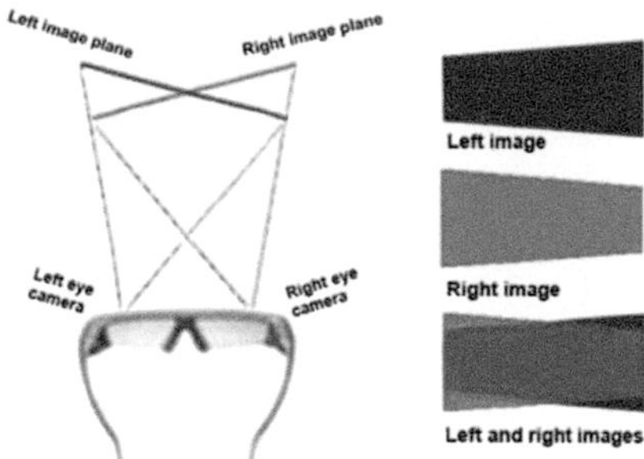

FIGURE 12.7: Converged shooting causes keystone artifacts. Figure from [60].

often it makes binocular fusion impossible. The solution is to create a "floating stereoscopic window," simply superimposing a black vertical band on the left of the left view and on the right of the right view. These bands also appear before the screen, hence the name, so with this extremely simple trick the occlusion is now consistent with the stereopsis and therefore the depth cue conflict disappears. This window should not be constant, though, because the audience would be distracted by *"this static window floating in front of them"* [180]. Dynamic floating windows are employed instead, and apart from solving the aforementioned depth cue conflict they are also used to simulate camera motion, moving the audience towards the scene or away from it [270].

12.5 Camera baseline and focal length

A change in the distance between the cameras, the camera baseline or interaxial distance, modifies the relative depth between objects in the scene; see Figure 12.8. And as we mentioned earlier, changing the focal length in regular 2D cinema affects the apparent distances between scene objects.

When the inter-axial distance is close to the average interocular distance and the focal length used by the cameras is close to the focal length of our eyes, the 3D objects appear natural. But otherwise, geometric 3D artifacts appear, visual problems that only arise when viewing the images in stereo and not each view separately. There are four types of these [233]:

- Cardboarding. When the cameras' focal length is larger than the eyes' focal length. Objects appear flat, as cardboard cut-outs.

- Pinching. When the cameras' focal length is shorter than the eyes' focal

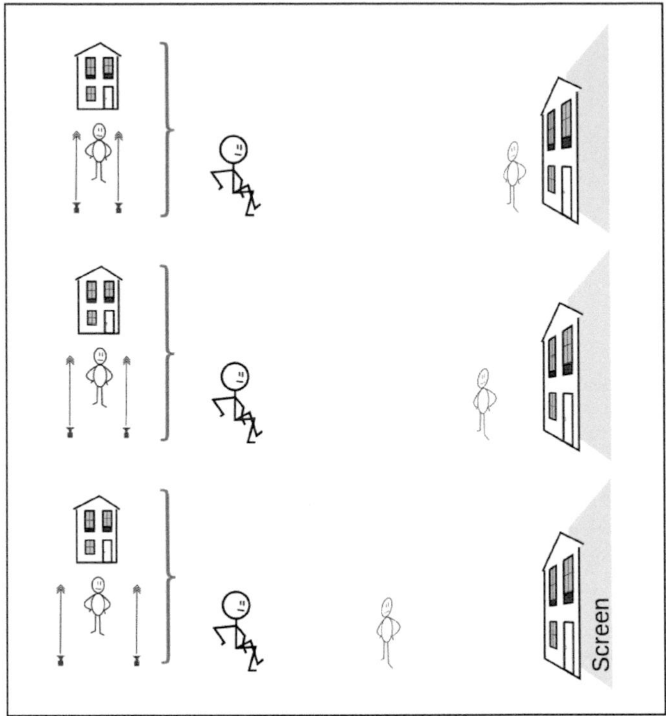

FIGURE 12.8: Modifying the inter-axial distance changes the relative depth. Figure adapted from [270].

length. Objects appear elongated or "pinched" along the direction of depth.

- Miniaturization. When the camera baseline is larger than the interocular distance. The object at the convergence point appears small, in a different scale from that of its surroundings.
- Gigantism. When the camera baseline is smaller than the interocular distance. The object at the convergence point appears large, in a different scale from that of its surroundings.

Pinching is less of a problem in practice because it's not that common to intensively use lenses much shorter than 15 mm. Cardboarding on the other hand is a real issue, because long lenses that are common in 2D cinema give poor results in S3D cinema, from 35 mm up, and with a focal length of 50 mm already producing "bad 3D" [270]. Not surprisingly, a focal length of 15 mm produces optimal results. In some situations, when the foreground and the background require different focal lengths, multiple camera rigs are used and the resulting footage must be composited in post-production.

Miniaturization and gigantism are determined by the baseline, and the baseline is quite difficult to alter in post-production, unlike the convergence. Altering the baseline requires generating a new view from the 3D scene, and this in turn implies that a very accurate depth map must be generated from the image disparities of stereo correspondences. A change in baseline may also be required in order to adapt for TV screens the depth range of an S3D movie shot for cinemas. We'll see how to do this in the following sections.

12.6 Estimating the depth map

The basis of all algorithms for depth map estimation is the principle of stereopsis, by which the disparity between corresponding images of the same scene point allows us to assess its depth: we associate large disparities with close points and small disparities with far away points. Depth map algorithms are therefore stereo matching algorithms, and they look for correspondences among the two views and impose regularity constraints in the solutions. For an overview and comparison of these types of techniques see [336]. Figure 12.9 compares the outputs of two methods with the actual ground truth of a standard (in computer vision) scene.

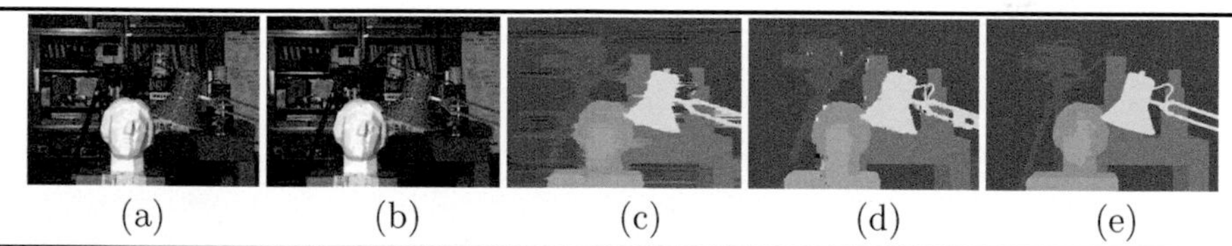

(a) (b) (c) (d) (e)

FIGURE 12.9: (a) Left view. (b) Right view. Depth maps: (c) estimated with dynamic programming, (d) estimated with graph cuts, (e) ground truth. Results from [336, 30].

12.7 Changing the baseline/synthesizing a new view

We have mentioned earlier that the need to synthesize a new view may arise in two contexts: when performing 2D-to-3D postconversion, and when we want to change the baseline of a native S3D movie in post-production. Having one view and its associated depth map, a new view can be generated simply by considering that the depth map values are proportional to corresponding pixel disparities. By shifting each pixel according to its value in the depth map, we create a new view; see Figure 12.10. Of course, at the boundaries between foreground and background objects the depth map is discontinuous, and these "jumps" provoke the appearance of image gaps in the synthesized view. For instance, if two pixels A and B are neighbors in the same line of the original view, and there is a jump in the depth value from A to B, then the corresponding pixels A' and B' in the new view will not be neighbors, the distance between them being proportional to the jump in depth values from A to B; furthermore, all the pixels in between A' and B' will be empty. These holes will have to be inpainted, as we discuss in Chapter 14.

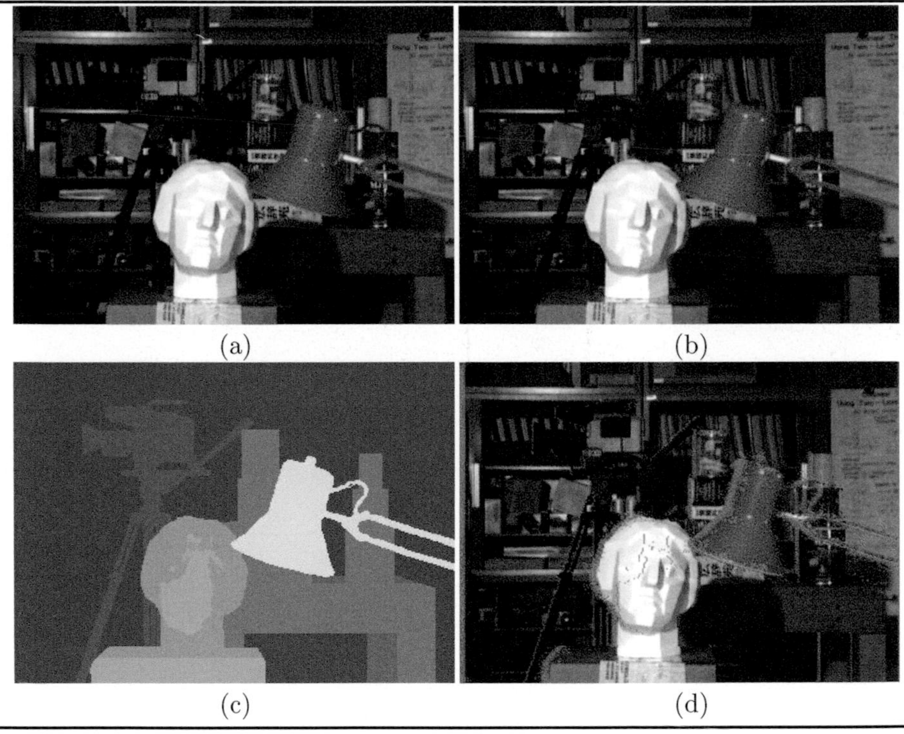

FIGURE 12.10: (a) Left view. (b) Right view. (c) Disparity map for (b). (d) New view synthesized from (b) by horizontally shifting each pixel according to its disparity value in (c), in green the disoccluded values that must be inpainted. This new view corresponds to reducing the baseline by half.

12.8 Changing the focus and the depth of field

Having a view and its associated depth map we can modify in post-production the focus point and the depth of field (DOF). For instance, Bertalmío et al. [97] propose a fast method where the original image is blurred and the amount of blur changes locally depending on the depth value. See examples in Figures 12.11 and 12.12.

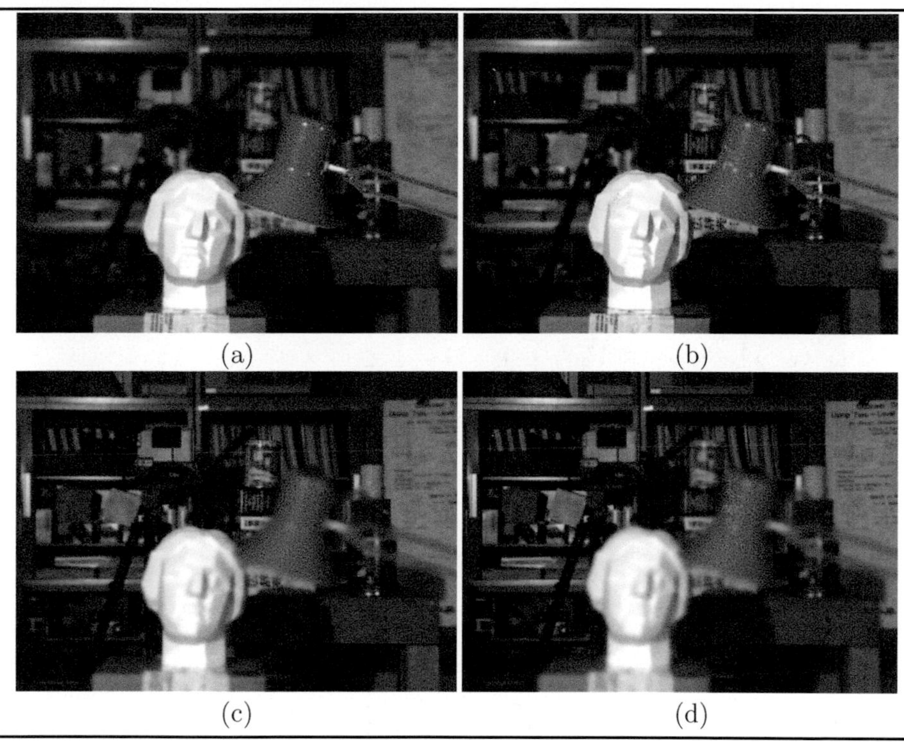

FIGURE 12.11: Changing the focus in post-production. (a) Focus on foreground lamp. (b) Focus on head sculpture. (c) Focus on table. (d) Focus on background. Results obtained with [97].

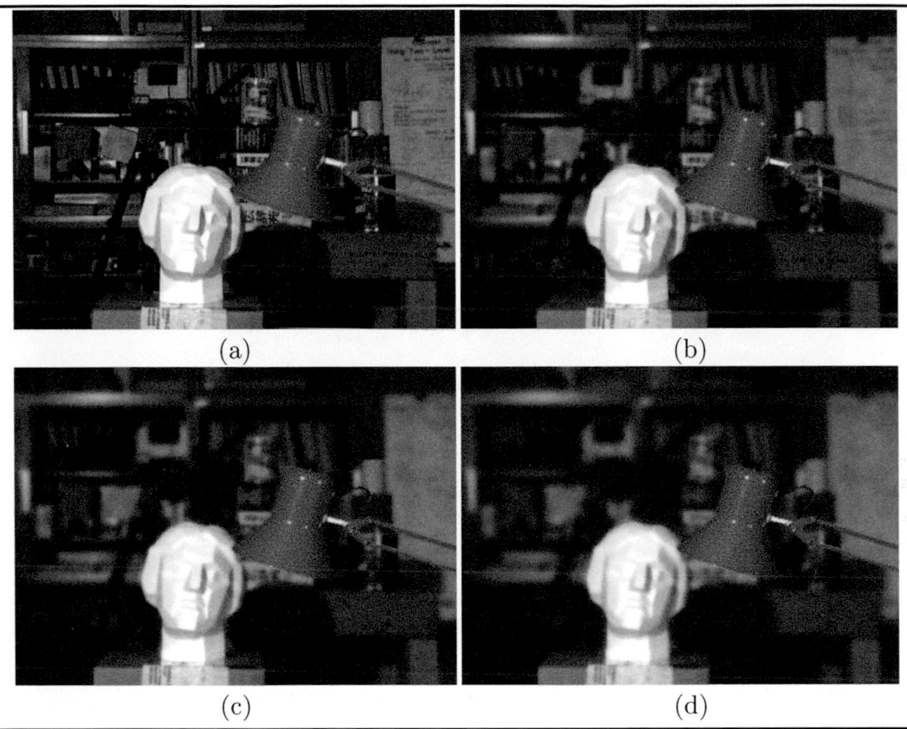

FIGURE 12.12: Changing the depth of field in post-production. The focus is always on the foreground lamp, but the depth of field diminishes from (a) to (d). Results obtained with [97].

12.9 Factors of visual discomfort

Visual discomfort is the subjective counterpart of visual fatigue (which refers to a decrease in performance of the human visual system), and there are several factors that are thought to produce it in stereoscopic cinema [241]:

- Excessive binocular disparity. As we mentioned regarding the extent of Panum's fusional area, fusion limits can be remarkably small. They decrease with smaller, detailed and stationary objects.
- Temporal changes in disparity.
- An artificial DOF. In everyday situations our eyes have a natural DOF and objects outside this range are blurry. In cinema, though, be it regular or S3D, the entire image is normally sharp, because different viewers may

concentrate on different parts of the screen. Therefore, sharp objects with disparity beyond the fusion limit still ellicit an effort for fusion in both eyes (because they are sharp), but they can't be fused because their disparity is too large. A possible solution would be to blur images according to their depth, as mentioned above in Section 12.8.

- Artifacts from 2D-to-3D conversion. With these techniques the depth is just approximate, because the image is partioned into an ordered sequence of (flat) layers. Spatial and temporal inconsistencies cause flicker or turbulence around the layers' edges. Dis-occluded parts (regions that become visible when shifting a layer to create a new view) must be inpainted, and this is a very challenging problem in the context of video.
- Blur facilitates depth perception, but it's an ambiguous cue (an object behind or before the focus is equally blurry). The human visual system doesn't integrate blur and binocular disparity because both cues are active over different ranges. But in stereoscopic images we are often forced to, and blur may favor an incorrect depth perception.

But many researchers argue that the main cause of visual discomfort is the accommodation and vergence mismatch. In real-life situations, the distance to objects is estimated from the linkage between accommodation and vergence. In a 3D display, on the other hand, the eyes are expected to converge on a virtual 3D object lying before or behind the screen, while at the same time the eyes are accommodating on the screen, where the image is sharpest (because as the distance between the virtual object and the screen increases, light becomes more diffuse and the object is perceived as blurry). Therefore, in stereoscopic 3D cinema the link between vergence and accommodation is *lost.* If the virtual object has too much disparity, the stereoscopic depth might be larger than the DOF and the result is impossibility of fusion (double vision), loss of accommodation (blur), or both. Walter Murch, one of the most influential film editors of all time, wrote a famous letter [3] to film critic Roger Ebert in which he argues that the accommodation and vergence mismatch is the reason why *"3D doesn't work and never will. Case closed."* It is an extremely well written argument that deserves to be reproduced here:

Hello Roger,

I read your review of "Green Hornet" and though I haven't seen the film, I agree with your comments about 3D.

The 3D image is dark, as you mentioned (about a camera stop darker) and small. Somehow the glasses "gather in" the image – even on a huge Imax screen – and make it seem half the scope of the same image when looked at without the glasses.

I edited one 3D film back in the 1980's – "Captain Eo" – and also noticed that horizontal movement will strobe much sooner in 3D than it does in 2D.

This was true then, and it is still true now. It has something to do with the amount of brain power dedicated to studying the edges of things. The more conscious we are of edges, the earlier strobing kicks in.

The biggest problem with 3D, though, is the "convergence/focus" issue. A couple of the other issues – darkness and "smallness" – are at least theoretically solvable. But the deeper problem is that the audience must focus their eyes at the plane of the screen – say it is 80 feet away. This is constant no matter what.

But their eyes must converge at perhaps 10 feet away, then 60 feet, then 120 feet, and so on, depending on what the illusion is. So 3D films require us to focus at one distance and converge at another. And 600 million years of evolution has never presented this problem before. All living things with eyes have always focussed and converged at the same point.

If we look at the salt shaker on the table, close to us, we focus at six feet and our eyeballs converge (tilt in) at six feet. Imagine the base of a triangle between your eyes and the apex of the triangle resting on the thing you are looking at. But then look out the window and you focus at sixty feet and converge also at sixty feet. That imaginary triangle has now "opened up" so that your lines of sight are almost – almost – parallel to each other.

We can do this. 3D films would not work if we couldn't. But it is like tapping your head and rubbing your stomach at the same time, difficult. So the "CPU" of our perceptual brain has to work extra hard, which is why after 20 minutes or so many people get headaches. They are doing something that 600 million years of evolution never prepared them for. This is a deep problem, which no amount of technical tweaking can fix. Nothing will fix it short of producing true "holographic" images.

Consequently, the editing of 3D films cannot be as rapid as for 2D films, because of this shifting of convergence: it takes a number of milliseconds for the brain/eye to "get" what the space of each shot is and adjust.

And lastly, the question of immersion. 3D films remind the audience that they are in a certain "perspective" relationship to the image. It is almost a Brechtian trick. Whereas if the film story has really gripped an audience they are "in" the picture in a kind of dreamlike "spaceless" space. So a good story will give you more dimensionality than you can ever cope with.

So: dark, small, stroby, headache inducing, alienating. And expensive. The question is: how long will it take people to realize and get fed up?

All best wishes,
Walter Murch

Chapter 13

Color matching for stereoscopic cinema

13.1 Introduction

As we have pointed out in Chapter 12, in professional 3D movie production it is crucial that the twin cameras used in a stereo set-up perform in exactly the same way, but color differences among views are common and often unavoidable, even in high-end productions. These problems are typically fixed in post-production by color-matching one view to the other, the master view, that is taken as reference.

In this chapter we reproduce excerpts and figures from the work by Bertalmío and Levine [98], which is a very simple yet very effective method for local color matching, consisting of three steps, each of them inspired by successful related works. Firstly, the master view is morphed so that it becomes locally registered with the source view, as Bertalmío and Levine very recently proposed in the context of image fusion for exposure bracketed pairs [99]. Next, color matching is performed on the source view by aligning each local histogram with the corresponding one in the warped target view; we will see that this histogram alignment is simply the mean shifting of the image values proposed by Reinhard et al. [324], although in a different color space. Finally, the Poisson editing technique of Pérez et al. [307] is applied on the source regions that have no correspondence in the target, so that the corrected colors are propagated inside these regions.

The contribution of [98] resides then in the novel combination of these existing methodologies so as to suit our particular problem, the simplicity of the overall approach (which has only one, very intuitive, parameter to set), its low computational cost, and the effectiveness of its results. This is demonstrated with examples from several professional, high quality 3D sequences that show that this algorithm outperforms four different popular color matching approaches.

13.2 Related work

In a recent work, Reinhard [323] surveys methods for color transfer among images, mentioning essentially four types of approaches: warping clusters or probability distributions in a three-dimensional color space, as in Pitié et al. [310]; matching means and standard deviations in an appropriate, decorrelated color space, as in Reinhard et al. [324]; aligning and scaling principal axes of the color distributions, as Kotera does in [234]; and straightforward histogram matching, performed independently on each color channel. The literature on the subject is extensive, and this is why we have decided to limit the comparisons of our algorithm to the four methods cited above. Incidentally, they are all *global* methods, i.e. they modify the value of each pixel ignoring its location and hence also ignoring the values of the neighboring pixels. Recent methods use pixel correspondences to identify similar regions between the image pair, but only to then refine the color matching transform which still is global, as in HaCohen et al. [190] and Kagarlitsky et al. [219]. Local color transfer methods do exist, such as the work of Tai et al. [352] where the images are segmented into regions in which color distributions are represented as Gaussian mixtures and then matched, or the work of Huang and Chen [202] where colors are transferred among corresponding landmarks and then extended to the rest of the image. While these methods provide very good results in a variety of situations, they are also much more involved than global approaches and rely on solutions to challenging tasks such as image segmentation or landmark selection and matching.

13.3 The algorithm

We will work in RGB and deal with each channel separately and in the same way. Because of this, in what follows we make a slight abuse of notation so that when we discuss images S and T we are referring to only one of their color channels. Likewise, histograms will always be a 1D histogram of a single color channel.

Our problem is to modify the source image S so that its colors locally match those of the target view T. We can do this in the following way: for each pixel $x \in S$,

- find its corresponding pixel $y \in T$: x and y are correspondents when they are the projections on S and T of the same 3D scene point;
- consider an $n \times n$ window $W_x \subset S$ centered at x, and find its corresponding region $W_y \subset T$;

- shift the histogram h_x of W_x so that it is aligned with the histogram h_y of W_y: this will give us a new value v' for $S(x)$;
- replace $S(x)$ with v'.
- in case we cannot find a correspondent y for x, often the case when x lies near a boundary of S which is not seen from the other view T, then $S(x)$ is not modified for now;
- after all possible pixels have been modified, their colors are propagated into the remaining, untouched pixels.

Each of these steps will now be described in detail, starting with a modification of the target image, which will allow us to increase the computational efficiency of our method without changing its behavior.

13.4 Morphing the target image

Since each W_x is square, we can compute the histograms h_x of S very efficiently: we can have stored in memory the current local histogram h_x so that when we move from x to its neighbor x', one pixel to the right, to compute $h_{x'}$ we only need to update h_x by adding n new values (a new column on the right) and removing n old values (corresponding to the left column). That is, computing the histogram of an $n \times n$ region in S requires only $2n$ look-up operations, not n^2. But notice that while W_x is always a square window by construction, its corresponding region W_y in general *will not* be square, and this poses practical problems for the efficiency of the computation of the histograms of T, which now go from linear to polynomial complexity.

This problem can easily be solved by making the following observation: with the map of all correspondences $x \leftrightarrow y$ between S and T we can morph T so that it's locally registered with S, i.e. deform T so that each pixel $y \in T$ moves to the position x but keeping the value $T(y)$ intact. This gives us a morphed target image T_m and for each pixel $x \in S$ its corresponding pixel $y \in T_m$ will be at the same location: $y = x$. Therefore, W_y will also be square and now we can compute efficiently the local histograms both at S and T_m.

The idea of morphing one image before comparing histograms was very recently proposed by Bertalmío and Levine [99] in the continuous setting: they have an energy functional with a term that measures the difference between a long-exposure blurry image and a short-exposure dark and noisy image, and they need to introduce "motion compensation" into this term in order to correctly match the colors of the functional minimizer to the colors of the blurry image. While we could think of applying that approach to our problem, the authors report a running time for their algorithm that is two orders of magnitude higher than ours, due to the iterative nature of their optimization procedure.

As for the morphing itself, we can use any technique that produces a dense map of correspondences between the two views. Given that they are stereo views there is abundant literature on the subject; see for instance the comprehensive evaluation of dense-stereo algorithms by Scharstein and Szeliski [336].

13.5 Aligning the histograms

In order to adapt the histogram h_x to the histogram h_y we have several options: straightforward histogram matching, dynamic time warping as Porikli does in [311], shifting so that some peak of h_x matches a corresponding peak of h_y, etc. We have tried the above solutions and finally decided upon the following one, for simplicity of computation and quality of results: treat the histograms as distributions of mass and shift h_x so that its barycenter matches the barycenter of h_y. The barycenter is a more robust measure than peak location, its computational cost is much smaller than that of dynamic time warping, and the subsequent shifting does not produce artifacts, unlike what happens with histogram matching.

But we can also make the following observations:

- each histogram value $m_i = h_x(r_i)$ represents the number m_i of pixels in the window W_x that have the value r_i;
- the barycenter G of h_x is computed as $G = \frac{1}{M}\sum_i m_i r_i$, where $M = \sum_i m_i$ is the total mass.

Therefore, G is just the average of all the window pixel values $S(q)$, $\forall q \in W_x$. Consequently, shifting h_x by matching barycenters simply amounts to matching the local means of image values in S and T_m. This is precisely the original approach for color transfer of Reinhard et al. [324], although they worked globally and in a different color space. Reinhard et al. used the variances as well to scale the values, but we have found that this is not necessary for our problem, where local color differences among stereo views of the same scene seem to be adequately modeled just by histogram shifts.

13.6 Propagating the colors to unmatched pixels

The above steps can only be applied to pixels x of S for which we have found a corresponding pixel y in T. But in stereo views there are always pixels that cannot be matched, typically due to occlusions or due to their proximity to an image border that makes them visible in only one view. Let Ω be the set of all pixels in S that cannot be matched, and $\partial\Omega$ the boundary of Ω. We have to color-match the pixels in Ω as well, otherwise our result would show visible color jumps at $\partial\Omega$. What we need then is to propagate the corrected colors of $\partial\Omega$ inside Ω, but maintaining all the image details and shapes of Ω intact.

This problem was posed and successfully solved in the Poisson editing work of Pérez et al. [307], by finding a solution for the Poisson equation in Ω with boundary conditions at $\partial\Omega$. The last step of our algorithm then is to apply Poisson editing to Ω so that finally all values of S are corrected. This process is usually very fast given that the area of Ω normally amounts to a very small percentage of the total area of S.

13.7 Examples and comparisons

Figure 13.1 shows example results obtained with our algorithm on stereo pairs from several professional sequences. Color differences are more noticeable in some examples than others, but all of them have large regions where the color mismatch goes well above 3%, a threshold value commonly used in film post-production as the maximum acceptable deviation in color. Notice for instance the colors on the hippo's cheek, the sky in the train image, the background trees and the foreground pavement in the car image, or the paint and cabin reflections in the plane image. Furthermore, several color mismatches appear to be local, as in the hippo and the plane images. With our method we are able to reduce all color differences and make most of them unnoticeable. The only parameter for our algorithm is the size n of the side of the windows W_x. In all cases we have used the same value of $n = 41$ (corresponding to a window half-width of 20) because it appears as a good compromise between computational efficiency and good sampling of the local histogram.

Figure 13.2 compares our method with four of the techniques mentioned in the introduction: Reinhard et al. [324], Pitié et al. [310], Kotera [234] and histogram matching. As we said the literature on the subject is very extensive, but these four works appear to be representative of the sort of approaches that are prevalent in color transfer. We can see that our method compares favorably in all four cases, but we must point out that our algorithm is specifically

FIGURE 13.1: Left column: original left view, to be color-corrected. Right column: original right view, considered as reference. Middle column: our result. Original images provided by Mammoth HD Footage library (www.mammothhd.com). Figure from [98].

adapted to the stereo color matching problem, while the other techniques are useful in many other situations (such as when the images are completely different, a case that cannot be handled with our approach).

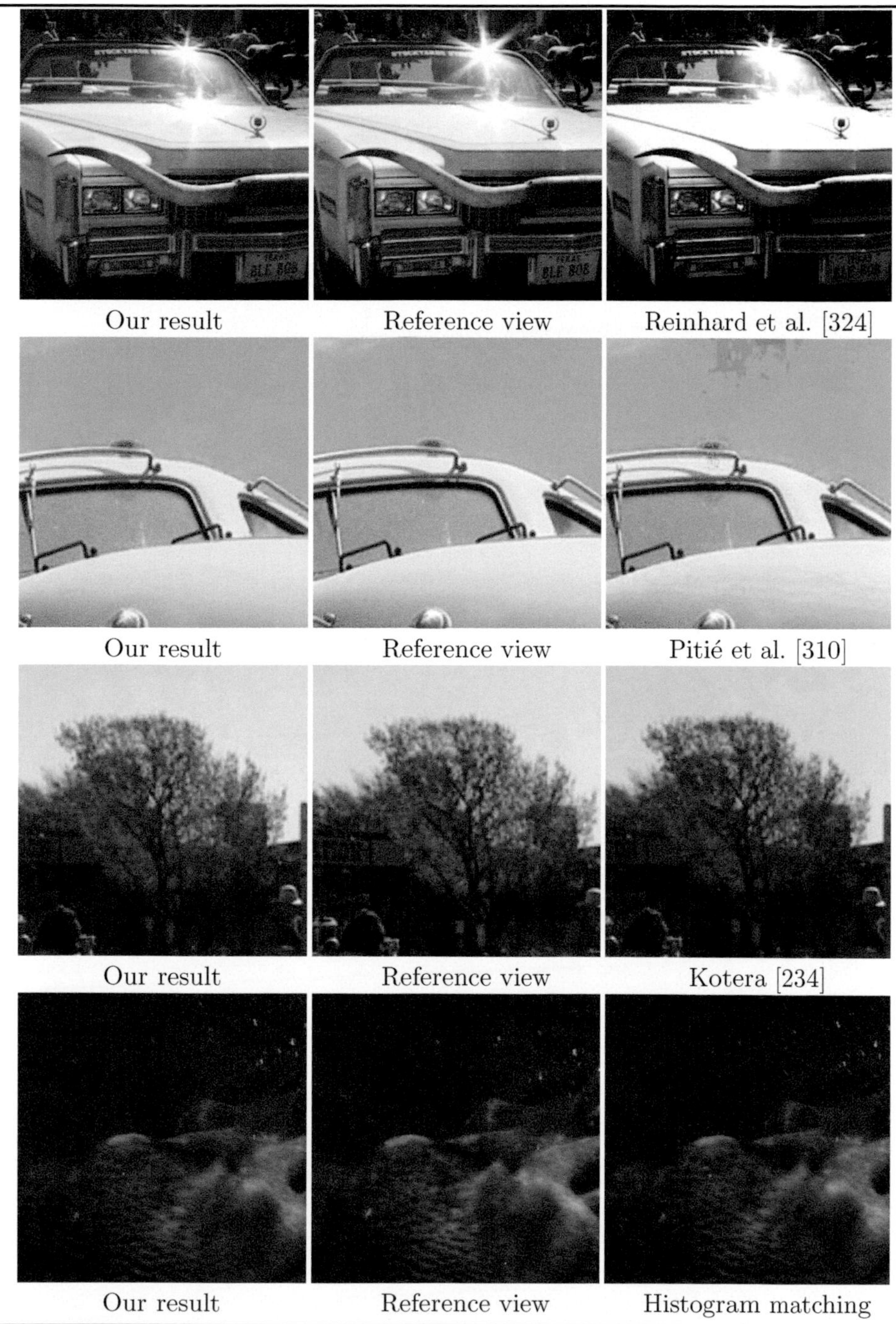

FIGURE 13.2: Details comparing the proposed method with the approaches of Reinhard et al. [324], Pitié et al. [310], Kotera [234] and histogram matching. Figure from [98].

Chapter 14

Inpainting

14.1 Introduction

The problem of automatically removing objects in video sequences is called *video inpainting*, as an extension of the *image inpainting* terminology that refers to the same problem for still digital images. This terminology was introduced in the year 2000 in the original "Image inpainting" article, by Bertalmío et al. [101], that adopted the denomination used in art restoration to denote the modification of a painting in a way that is non-detectable for an observer who does not know the original artwork. Other popular terms are *image completion*, *image fill-in*, *image repair* and *image disocclusion*, the latter being introduced by Masnou and Morel [265] in their pioneering approach, which treated the image hole or gap as an occluding object that had to be removed. Image denoising algorithms do not apply to image inpainting, because all the original information present in the image gap is discarded. For a recent overview on inpainting see [92].

In the cinema industry, the need for inpainting may arise because of several reasons, such as:

- there are objects appearing in frame that must be removed, e.g. a microphone, a brand logo, a present-day object in a period piece, etc.;
- after view synthesis in S3D cinema some regions that were occluded in the original view now become visible, but there's no image information for them and this information must be inpainted (see Chapter 12);
- after camera stabilization there are missing regions at the frame boundaries (see Chapter 8).

The video inpainting problem consists, then, in, given a sequence with T images I^n, $0 \leq n < T$, where each image I^n has a marked region Ω^n that should be removed, how to obtain a restored sequence R^n where $R^n(i,j) = I^n(i,j)$ if pixel (i,j) lies outside Ω^n, but if (i,j) is inside Ω^n it is assigned a color so that it harmonizes with its surroundings and the elimination of object Ω^n is not perceptible to anyone not knowing the original sequence. Figure 14.1 shows an example.

All inpainting algorithms, both for still images and video, assume that the masks or set of masks are given: obtaining these masks is a problem in itself, which in practice and for image inpainting is solved manually (the user just creates a binary mask by painting over the region to be removed) but that for video is automated with segmentation and tracking algorithms (*rotoscoping* in the movie industry jargon).

FIGURE 14.1: Top row: some frames from a video. Middle row: inpainting mask Ω in black. Bottom row: video inpainting results obtained with the algorithm of Patwardhan et al. [305].

For still images ($T = 1$) there is abundant literature, with algorithms that can be categorized in [113, 92]:

- Geometrical methods (using variational principles and Partial Differential Equations (PDEs)). The image gap is filled in by imposing continuity and smoothness on the image level curves (as in [101, 91, 90]) or by minimizing an energy functional (as in [265, 85]). In [301] there is a survey on these sorts of approaches, which are well suited for piecewise smooth images or when the gap Ω is thin in comparison with the image structures that surround it.

- Texture synthesis methods, which search outside Ω for blocks to copy and paste into Ω. This was proposed in the seminal work of Efros and Leung [153], which not only spawned a large class of very successful "exemplar-based" inpainting techniques, but also inspired the Non-Local Means algorithm of Buades et al. [109] that revolutionized the image denoising community [246], as we saw in Chapter 5.

- Compact representation methods. The image is inpainted by expressing it in the most compact way using elements from an image dictionary [350, 329, 154, 351].

Several works (e.g. [102, 321, 133, 113]) argue that the best performing algorithms are those that combine elements of two or the whole three of the aforementioned types of approaches. Each image inpainting algorithm has its advantages and shortcomings, but a trivial extension from still images to video never works. If we treat the video sequence as a set of T images that are inpainted independently from each other using any technique for still images, then each output frame of the inpainted sequence may be of sufficient quality but when the movie is played there will appear very noticeable visual artifacts. These are due to not having taken into account the temporal consistency of the video when performing the inpainting process, and our visual system is extremely sensitive to temporal inconsistencies.

14.2 Video inpainting for specific problems

The development of image inpainting techniques became a very active research area from 2000 onwards, but previously there had appeared solutions for specific video inpainting problems.

Basic methods for video inpainting for film restoration applications (dealing with image gaps produced by dust, scratches or the abrasion of the material) assume that the missing data changes location in correlative frames, and therefore use motion estimation to copy information along pixel trajectories. A particular difficulty in video inpainting for film restoration is that, for good visual quality of the outputs, the detection of the gap and its filling in are to be tackled jointly and in a way that is robust to noise, usually employing probabilistic models in a Bayesian framework; see for example the book by Kokaram [232].

In data transmission the problem is known as "error concealment." Data blocks that have been lost during transmission can be recovered thanks to the high temporal redundancy of the video information. Typically, the motion vectors of the lost blocks are estimated and used to find the corresponding full blocks previously decoded [119, 362, 240, 70].

Both types of techniques can only be applied successfully on sequences where the image gaps Ω are small, don't last for very many consecutive frames, and have a predetermined shape. All these limitations cannot be assumed in a general setting of the video inpainting problem.

14.3 Video inpainting in a general setting

The techniques that we'll cover in this section are presented in their respective articles as general video inpainting methods, but they all assume a series of limitations that are common to most of them and that make the video inpainting problem still pretty much an open one. These limitations refer to camera motion (e.g. that it is non-existent, or that it is parallel to the image plane), to the motion of the region that is to be inpainted (e.g. that it must be periodic, without change of scale or perspective), to the scene background (which must be static or have a periodic motion at most), etc. These restrictions are satisfied by a large number of sequences, of course, but it is also true that most "natural" or real video sequences violate some or all of these assumptions to some extent. Now we will describe some of the more relevant video inpainting techniques.

The video inpainting method of Wexler et al. [379] considers the video input as a spatiotemporal volume in which a 3D extension of the texture synthesis technique of Efros and Leung [153] is applied: instead of matching patches they match cubes, and the values to compare are not only the RGB triplets of each pixel but also its 2D motion vector. After all the best matches have been found, the final output value for each pixel is computed as the average of the overlapping matching cubes at that position: this is based on the idea of coherence among neighbors developed by Ashikhmin [75], and was later incorporated as a key element in state-of-the-art denoising techniques under the name "aggregation" (see [246] and Chapter 5). Inpainting is tackled then as a global optimization problem, therefore the computational cost is very significant. A multi-scale approach is introduced to speed up the process, starting the search for correspondences at the coarsest level and refining the results as they go upwards on the image pyramid. It is implicitly assumed that objects move in a periodic way and that there aren't any changes in scale or perspective: otherwise, the simple "copy and paste" approach of [153] on which this method is based would not work. For each pixel, the best candidates are averaged (instead of randomly choosing one as it was done in [153]), and this causes a quite noticeable blur in the final result. In the examples shown in [378, 379] the camera is always static, probably due to the fact that for the global optimization procedure a very simple motion estimation technique is used: this technique approximates time derivatives with central differences in a $3x3x3$ stencil, so this estimation is very sensitive to object displacements in the image.

Cheung et al. [121] propose a method based in a probabilistic modeling of the video. *Video epitomes* are defined as probabilistic models based in cubes (3D blocks in video). Thes models are learnt from a database with a large number of example cubes taken from natural image sequences. Just one parametric form is used, a 3D Gaussian parameterized with different means and covariance matrix for each entry. These video epitomes are then employed to synthesize data in the areas of interest, the image gaps Ω. Each pixel wil have as predictors several overlapping cubes, hence the final result will be a weighted average. Results are very similar to those of [378], with low resolution examples and excessive blur.

Jia et al. [212, 211] propose a very involved video inpainting technique that combines several methods and also requires extensive user interaction, since the user must manually select the boundaries of the different depth layers into which the video sequence is decomposed. The algorithm must "learn" the statistics of the background (the layer of greater depth). Object motion must be periodic, avoiding changes in scale and perspective. The camera must be static or move laterally in a plane parallel to the image plane. *Tensor Voting* [210] is used inside each layer to segment textures and later to synthesize them. Results are good but they are not exempt from artifacts. If a moving object is partially occluded by the object that we wish to remove, then the moving object is synthesized from scratch, superimpiosed over the sequence and moved along a novel trajectory. This procedure causes very noticeable artifacts, because objects move in an unrealistic manner (e.g. a walking person appears at some point to be floating above the ground). A related method, which also combines depth layer estimation and segmentation with warping and filling in, is that of Zhang et al. [402].

Jia et al. [392] propose a method based in texture synthesis. Once again, this implies the assumption that objects move in a periodic fashion without changing scale or perspective. The authors use tracking to reduce the size of the search space, and the minimization technique of Graph Cuts to combine synthesized blocks. This approach can only be used on static-camera sequences. The authors do not provide video examples, but they report that results suffer from artifacts near the boundaries of the image gaps, and also that the inpainting process may fail if tracking is lost.

Shiratori et al. [347] apply the texture synthesis technique of Efros and Leung [153] to the field of motion vectors. Once the field has been restored inside the gaps, the motion trajectories for the whole video are complete and the missing colors are obained through bilinear interpolation along the trajectories. The main shortcomings of the approach are the blurring caused by the interpolation method and the limitation to sequences where motion information in enough to fill in the holes, which has the implicit assumption that the information to be restored is fully present in the video sequence and this is not always the case (for instance if the hole and the camera are static). A similar method with the same limitations is proposed by Zhao et al. [404], where the optical flow inside Ω is reconstructed first and then this information is used to restore the colors, in the spirit of the "error concealment" methods mentioned before for data transmission.

Patwardhan et al. [305] propose a video inpainting method that combines several techniques but that in essence is extending to video the non-parametric sampling method of Criminisi et al. [133], which is itself a combination of the texture synthesis technique of Efros and Leung [153] and the edge-propagation inpainting of Bertalmío et al. [101]. In the first step they decompose the video sequence into binary motion layers (foreground and background), which are used to build three image *mosaics* (a mosaic is the equivalent of a panorama image created by stitching together several images): one mosaic for the foreground, another for the background and a third for the motion information. The other two steps of the algorithm perform inpainting, first from the foreground and then from the background: these inpainting processes are aided and sped up by using the mosaics computed in the first step. See Figure 14.1 for some results. The algorithm is limited to sequences where the camera motion is approximately parallel to the image plane, and foreground objects move in a repetitive fashion and do not change size: these restrictions are imposed in order to use the patch-synthesis algorithm of Efros and Leung [153], which can't handle changes in perspective or scale. Also, the method in [305] is not able to deal with complete occlusions of moving objects, nor with changes in illumination, as in [212].

Shen et al. [343] start by warping the spatio-temporal 3D video volume so that each frame is registered with the first and the resulting volume is equivalent to the case of a static camera. After foreground-background separation, the foreground volume is rectified so that its global trajectory motion is parallel to the XT plane (horizontal-temporal plane); this involves assuming that the motion is cyclic as well as manually selecting some corresponding points. Then the trajectory of the object is obtained in this XT slice and used to sample patches near this path that will be used for texture synthesis through an energy minimization process. The same technique is used for repairing the background, but in this case volume rectification isn't necessary. Remaining holes are filled in in the first frame with the approach by Criminisi et al. [133], then copied to all other frames.

Shah et al. [71] propose a method consisting of minimizing a functional that measures the regularity of a vector field similar to the optical flow, constraining image transformations to being affine and finally estimating the parameters of some spline curves that approximate trajectories in the restored sequence. Perspective changes are limited, because of the use of affine transformations to approximate projective transformations. The examples shown in the paper seem to indicate that the method is not able to handle occlusions too well. For moving objects, the article mentions that the motion vector field should be segmented but they do not detail how to perform this operation. The camera is static in all the examples in the document.

Venkatesh et al. [367] focus on inpainting human motion in a short time window, and also restrict the camera motion to being parallel to the image plane. They follow [305] in performing foreground-background separation and creating a background panorama. The segmented foreground objects are treated as object templates and stored in a database. Background inpainting is performed with the method of Criminisi et al. [133], while the partially or totally occluded foreground human subjects are restored by combining patches from the template database. The results show visible blur and lightness discontinuities.

The main contribution of the work by Shih et al. [346] is to allow for camera motions such as zooming and panning. Inpainting on the first frame is performed with the priority-based exemplar scheme of Criminisi et al. [133], then propagated to the next frame with a transformation that is defined as a translation in the paper (a shift by the average motion vector of the surroundings), but that the authors claim can be extended, although not easily, to cope with rotations and perspective changes. The remaing portions of the hole in the next frame are filled with the method of [133] as well, and the whole process is iterated, from one frame to the next. One of the main limitations of this technique is its inability to perform accurate patch matching on scaled and/or rotated objects.

Tang et al. [356] propose a video inpainting technique that is tailored to the restoration of severely damaged old films. They perform motion field completion in a way similar to that of Shiratori et al. [347], then use the restored motion vectors to assist the patch matching which is based on the method of Criminisi et al. [133], and finally apply the Poisson editing technique of Pérez et al. [307] to blend the selected matches into the damaged areas. The main limitations of this method are its reliance on accurate motion estimation and intermediate frame completion results and its inability to deal with gaps of large area and zoom in-out camera motion.

Ebdelli et al. [149] extend the video inpainting algorithm of Patwardhan et al. [305]. For each pixel and corresponding image patch centered in it, they find the K most similar patches according to the usual Sum of Squared Differences (SSD) measure. These K patches are combined in two different ways using neighbor embedding techniques, instead of just picking the most similar patch, as in [305]. From these two synthesized patches, the one kept for the final output is the one having the lowest SSD with respect to the known pixels of the input patch. The authors mention that their algorithm suffers from error propagation due to uncertainties on the estimated motion information.

Granados et al. [187] propose a method that allows for both arbitrary camera and object motion, and does not require dense depth estimation or 3D reconstruction. They assume that the object to be removed occludes static background, and that each point in the hole is seen in at least one frame. They approximate the scene by an arrangement of planes, which lets them align frame pairs by matching image regions through homographies. This procedure gives several possible matches for each pixel; a single match is chosen by minimizing an energy measuring the consistency of the match candidate's frame with its neighbors. Since neighboring pixels may be filled with values coming from frames far apart in the video sequence, lighting differences are prone to appear and they are reduced with a Poisson blending technique like the one of Pérez et al. [307], as in Tang et al. [356]. The main limitation mentioned by the authors is the appearance of temporal artifacts, which is to be expected since the algorithm does not impose temporal consistency; they argue that temporal coherence would yield a restrictive limitation in the number of possible sources, which would lead to poorer results.

The method by Mosleh et al. [288] takes an approach essentially very similar to the one of Patwardhan et al. [305], but the patch matching of RGB values and motion vectors is replaced by a patch matching in the bandlet coefficient domain (the bandlet transform constructs orthogonal vectors that are elongated in the direction of image contours.)

14.4 Video inpainting for stereoscopic 3D cinema

In the context of stereoscopic 3D (S3D) cinema, the need for video inpainting appears in two different scenarios:

1. There is an object to remove in both views of an S3D movie.

2. From the two stereo views (or from just one view and its correspondent *depth map*), a new view has to be synthesized that will present gaps corresponding to occluded pixels that become visible in this view, and these gaps have to be inpainted.

The literature for the first problem is quite limited, while in the second case there are more works but many are geared towards a real-time application in 3D TV displays, therefore tending to compromise quality for speed. We will now briefly discuss the most relevant works in both scenarios.

14.4.1 Stereoscopic inpainting

Wang et al. [370] propose an algorithm for stereoscopic inpainting of *images*, not video. The inputs are: a stereo pair, their corresponding disparity maps (computed with some stereo matching algorithm, as seen in Chapter 12), the occluded areas in both maps and the masks marking the hole regions in both views. First the disparity maps are inpainted inside the occluded areas: the maps are segmented, each segment approximated by a plane, and each occluded pixel assigned a segment (and therefore a depth value, according to the plane equation associated with the segment) by minimizing an energy that penalizes color differences, spatial distance between segments and inconsistent visibility relationships (i.e. when a pixel is occluded by another that is actually further away, not closer). Next, the images are warped and the image gaps completed as much as possible. Then the rest of the pixels in the image gaps are filled in by an adaptation of the exemplar-based method of Criminisi et al. [133], which now considers not only RGB values but also the disparities (depth values). This texture synthesis is performed independently for each view, so a last step is required to impose consistency: if the patches surrounding two corresponding pixels are different above a threshold, the inpainted pixels are discarded.

Hervieu et al. [195] compute and inpaint the disparities simultaneously, and the diffusion of the depth values is performed using a total variation (TV) approach and applied on the image gaps, not just the occluded pixels as in the previous method. These restored disparity maps are used to fill in portions of the gap in one view that may be visible in the other view. Then an exemplar-based texture synthesis algorithm is applied to both views at the same time: the algorithm of Criminisi et al. [133] is adapted so that when a patch is used for inpainting in one view, it is warped (according to the disparities) and used to fill in the other view. This way the consistency among views is enforced.

Morse et al. [287] take a similar overall approach as the two abovementioned methods: first the disparity maps are inpainted, in this case with the PDE-based method of Bertalmío et al. [101], then a texture synthesis technique is applied to both views, in this case a stereo-consistent modification of the PatchMatch approach of Barnes et al. [87]. Finally, the found patches are blended, giving more weight to those patches that are consistent with their corresponding patch in the other view. Figure 14.2 compares the results of [287] and [370].

FIGURE 14.2: First two columnns: original images. Columns 3 and 4: stereoscopic completion by Morse et al. [287]. Last column: one view of the stereoscopic completion of Wang et al. [370]. Figure from [287].

Raimbault and Kokaram [318] propose a method for stereo *video* inpainting that esentially applies the same ideas of the methods for stereo image inpainting just mentioned (filling in of disparities, followed by texture synthesis) only extended to video: now the motion vectors are also considered, and the exemplar-based method of Criminisi et al. [133] is adapted so as to look for matches in different frames. This texture synthesis technique is applied to one view and then to the other, using pixel values that have already been filled in. For patch matching, a distance based on structural information is used instead of the usual SSD. The authors mention several limitations of their algorithm, including the need for manual selection of parameters, the reliance of the results on the accurate estimation of motion and disparity vectors, the difficulties in handling occlusions and disocclusions and the problems in recovering complex backgrounds. This method was soon improved by Raimbault et al. [319] by introducing several changes: processing one frame and immediately afterwards its stereo-counterpart, enforcing stereo consistency in the distance used to compare patches, and copying whole (or partial) patches instead of single pixels. The authors report that this method performs better, but it still has problems handling camera tilt, zoom and perspective.

14.4.2 Inpainting occlusions in stereo views

We saw in Chapter 12 how the inter-ocular distance in S3D movies had to be different in cinemas than in TV displays. This implies that a movie shot in S3D for cinema release has to undergo another 3D postproduction process for TV or BluRay release, in which a new view has to be synthesized. From the two original views a depth map is computed, and this depth map is used to warp one of the views to its new intended position. This warping is just a pixel-wise horizontal shift proportional to the disparity value in the depth map, so wherever there is a discontinuity in the depth map (e.g. at the boundaries of foreground objects) there will be an empty region in the warped image, that has to be inpainted. The same problem happens in 2D to 3D conversion, where only one view is available, a depth map is created by an artist, and from these two elements another view is synthesized. The generic name for the techniques that deal with this problem, of taking a view and a depth map and interpolating one or two views and filling in gaps, is *Depth Image Based Rendering* (DIBR). In these cases, where the top priority is the quality of the results and the computations are performed off-line, a regular (not stereoscopic) video inpainting method like the ones presented above might be adapted and employed for this specific problem, where apart from the video input we have the depth map: Schmeing and Jiang [338], Ndjiki-Nya [294] and Choi et al. [126] propose variations on the method of Patwardhan et al. [305], Cheng et al. [120] adapt the non-parametric sampling method of Criminisi et al. [133] with depth-aware patch comparisons in space and time, and Oh et al. [296] perform depth-aware geometric inpainting.

Another instance of the same problem happens with S3D movies in "view plus depth" format, where instead of coding the two views only one view and its depth map are represented (in this way the compression ratio that can be achieved is higher) and the display or the disc player has to generate the two views in real time and inpaint the disocclusions. For such applications, fast methods that regularize the depth-map or just interpolate it with simple aproaches are preferred (see references in [126]), but their obvious drawback lies in the poor visual quality of the results, where the scene geometry has been modified and temporal consistency is not taken into account.

14.5 Final remarks

The inpainting problem on still images became a popular research subject after the year 2000, and the literature on the topic is vast and still growing. But the extension of these techniques to video isn't in any way straightforward, because applying image inpainting techniques to each frame independently causes very noticeable visual artifacts, and achieving temporal consistency is actually quite hard. This is why the video inpainting literature is scarce, with the first significant works dating from the mid 2000s. And even so, these video inpainting techniques have a very limiting set of (implicit or explicitly stated) assumptions that make them unsuitable for many video sequences. Video inpainting remains then very much an open problem, and until there is a significant breakthrough in the field the process of removing unwanted objects from films will continue to be a software-assisted but mostly manual one in the movie industry.

Bibliography

[1] Online edition of Brian Wandell's *foundations of vision*: http://foundationsofvision.stanford.edu/.

[2] http://blog.abelcine.com/2012/11/01/sonys-pmw-f5-and-f55-defining-cfa/.

[3] http://blogs.suntimes.com/ebert/2011/01/post_4.html.

[4] http://digital.warwick.ac.uk/goHDR/.

[5] http://dpanswers.com/content/tech_defects.php.

[6] http://electronics.howstuffworks.com/camera1.htm.

[7] http://en.wikipedia.org/wiki/3D_film.

[8] http://en.wikipedia.org/wiki/Afocal_system.

[9] http://en.wikipedia.org/wiki/Black-body_radiation.

[10] http://en.wikipedia.org/wiki/Checker_shadow_illusion.

[11] http://en.wikipedia.org/wiki/CIELUV.

[12] http://en.wikipedia.org/wiki/Color_gamut.

[13] http://en.wikipedia.org/wiki/Color_matching_function#Color_matching_functions.

[14] http://en.wikipedia.org/wiki/Depth_perception.

[15] http://en.wikipedia.org/wiki/Dichroic_prism.

[16] http://en.wikipedia.org/wiki/Discrete_cosine_transform.

[17] http://en.wikipedia.org/wiki/Isaac_Newton.

[18] http://en.wikipedia.org/wiki/JPEG.

[19] http://en.wikipedia.org/wiki/Spectral_sensitivity.

[20] http://en.wikipedia.org/wiki/Standard_illuminant.

[21] http://en.wikipedia.org/wiki/Zoom_lens.

[22] http://hypertextbook.com/facts/2002/JuliaKhutoretskaya.shtml.

[23] http://johnbrawley.wordpress.com/2012/09/03/in-the-raw/.

[24] http://pages.videotron.com/graxx/CIE2.html.

[25] http://reviews.cnet.com/8301-33199_7-57437344-221/active-3d-vs-passive-3d-whats-better/.

[26] http://scubageek.com/articles/field_of_focus.pdf.

[27] http://scubageek.com/articles/wwwfnum.html.

[28] https://www.youtube.com/editor.

[29] http://variety.com/2013/digital/features/summerfx-3d-conversion-1200485853/.

[30] http://vision.middlebury.edu/stereo/. Middlebury Stereo Vision Page, StereoMatcher program.

[31] http://www.adamwilt.com/TechDiffs/Sharpness.html.

[32] http://www.brucelindbloom.com/index.html?Eqn_RGB_XYZ_Matrix.html.

[33] http://www.cambridgeincolour.com/tutorials/lens-quality-mtf-resolution.htm.

[34] http://www.dcimovies.com/specification/index.html.

[35] http://www.dr-lex.be/info-stuff/3dfail.html.

[36] http://www.eoshd.com/content/9148/.

[37] http://www.extremetech.com/extreme/124965-titanic-how-do-you-convert-a-movie-to-3d-anyway.

[38] http://www.extremetech.com/extreme/145168-3d-tv-is-dead.

[39] http://www.flickr.com/photos/-adam.

[40] http://www.flickr.com/photos/27718575@N07.

[41] http://www.flickr.com/photos/avidday.

[42] http://www.flickr.com/photos/damianathegirl.

[43] http://www.flickr.com/photos/dennistt.

[44] http://www.flickr.com/photos/fhke.

[45] http://www.flickr.com/photos/jpstanley.

[46] http://www.flickr.com/photos/mescon.

[47] http://www.flickr.com/photos/nickbedford.

[48] http://www.flickr.com/photos/pgoyette.

[49] http://www.flickr.com/photos/robertsmith.

[50] http://www.flickr.com/photos/shareski.

[51] http://www.flickr.com/photos/torontocitylife.

[52] http://www.flickr.com/photos/visitportugal/.

[53] http://www.flickr.com/photos/vox.

[54] http://www.flickr.com/photos/xparxy.

[55] http://www.flickr.com/photos/yakobusan.

[56] http://www.flickr.com/photos/zengei.

[57] http://www.freeflysystems.com/products/moviM10.php.

[58] http://www.guardian.co.uk/film/2012/mar/06/3d-films-lose-appeal.

[59] http://www.handprint.com/HP/WCL/color2.html.

[60] http://www.lightillusion.com/stereo_3d.html.

[61] http://www.math.ubc.ca/~cass/courses/m309-01a/chu/intro.htm.

[62] http://www.mpi-inf.mpg.de/resources/hdr/gallery.html. MPI Database.

[63] http://www.red.com/learn/red-101/hdrx-high-dynamic-range-video. HDR technology for RED cameras.

[64] http://www.red.com/learn/red-101/resolution-aliasing-motion-capture.

[65] http://www.theregister.co.uk/2013/07/05/bbc_cuts_3d_tv_service/.

[66] www.cis.rit.edu/fairchild/HDRPS/HDRthumbs.html. Fairchild Database.

[67] J. Adams, K. Parulski, and K. Spaulding. Color processing in digital cameras. *Micro, IEEE*, 18(6):20–30, 1998.

[68] J.E. Adams Jr, A.T. Deever, E.O. Morales, and B.H. Pillman. Perceptually based image processing algorithm design. *Perceptual Digital Imaging: Methods and Applications*, 6:123, 2012.

[69] G. Adcock. Charting your camera. *Creative COW Magazine*, November 2011.

[70] M.E. Al-Mualla, N. Canagarajah, and D.R. Bull. Motion field interpolation for temporal error concealment. *EE Proceedings-Vision, Image and Signal Processing*, 2000.

[71] O. Alatas, P. Yan, and M. Shah. Spatiotemporal regularity flow (spref): its estimation and applications. *Circuits and systems for video technology, IEEE Transactions on*, 2007.

[72] A. Alsam and I. Farup. *Computational Color Imaging*, chapter Spatial colour gamut mapping by means of anisotropic diffusion, pages 113–124. Springer Berlin Heidelberg, 2011.

[73] A. Alsam and I. Farup. Spatial colour gamut mapping by orthogonal projection of gradients onto constant hue lines. In *Proceedings of 8th International Symposium on Visual Computing*, pages 556–565, 2012.

[74] L.E. Arend Jr, A. Reeves, J. Schirillo, R. Goldstein, et al. Simultaneous color constancy: papers with diverse Munsell values. *Journal of the Optical Society of America A*, 8(4):661–672, 1991.

[75] M. Ashikhmin. Synthesizing natural textures. *Proceedings ACM Symp. Interactive 3D Graphics, ACM Press*, pages 217–226, 2001.

[76] M. Ashikhmin. A tone mapping algorithm for high contrast images. In *Eurographics Workshop on Rendering*, pages 1–11. P. Debevec and S. Gibson Eds., 2002.

[77] F. Attneave. Some informational aspects of visual perception. *Psychological review*, 61(3):183, 1954.

[78] (No author). How to measure MTF and other properties of lenses. Technical report, Optikos Corporation, 1999.

[79] S.P. Awate and R.T. Whitaker. Higher-order image statistics for unsupervised, information-theoretic, adaptive, image filtering. In *IEEE Computer Society Conference on Computer Vision and Pattern Recognition*, volume 2, pages 44–51. IEEE, 2005.

[80] S.P. Awate and R.T. Whitaker. Unsupervised, information-theoretic, adaptive image filtering for image restoration. *IEEE Transactions on Pattern Analysis and Machine Intelligence*, pages 364–376, 2006.

[81] T.O. Aydin, R. Mantiuk, K. Myszkowski, and H.P. Seidel. Dynamic range independent image quality assessment. *ACM Transactions on Graphics*, 27(3), 2008.

[82] A. Ayvaci, H. Jin, Z. Lin, S. Cohen, and S. Soatto. Video upscaling via spatio-temporal self-similarity. In *21st International Conference on Pattern Recognition (ICPR)*, pages 2190–2193. IEEE, 2012.

[83] S. Baker, E. Bennett, S.B. Kang, and R. Szeliski. Removing rolling shutter wobble. In *IEEE Conference on Computer Vision and Pattern Recognition (CVPR)*, pages 2392–2399. IEEE, 2010.

[84] C. Ballester, M. Bertalmío, V. Caselles, L. Garrido, A. Marques, and F. Ranchin. An Inpainting-Based Deinterlacing Method. *IEEE Trans. Image Processing*, 16(10):2476–2491, 2007.

[85] C. Ballester, M. Bertalmío, V. Caselles, G. Sapiro, and J. Verdera. Filling-in by joint interpolation of vector fields and grey levels. *IEEE Transactions on Image Processing*, 10, 2001.

[86] D. Bankston. The color-space conundrum. *American Cinematographer*, 86(4), 2005.

[87] C. Barnes, E. Shechtman, A. Finkelstein, and D.B. Goldman. Patchmatch: A randomized correspondence algorithm for structural image editing. *ACM Transactions on Graphics*, 28(3):2, 2009.

[88] S. Battiato, G. Messina, and A. Castorina. *Single-Sensor Imaging: Methods and Applications for Digital Cameras*, chapter Exposure correction for imaging devices: an overview, pages 323–349. CRC Press, 2008.

[89] A. Bazhyna. Image compression in digital cameras. *Tampereen teknillinen yliopisto, Julkaisu-Tampere University of Technology*, 2009.

[90] M. Bertalmío. Strong-continuation, contrast-invariant inpainting with a third-order optimal pde. *IEEE Transactions on Image Processing*, 15(7):1934–1938, July 2006.

[91] M. Bertalmío, A. Bertozzi, and G. Sapiro. Navier-Stokes, fluid-dynamics, and image and video inpainting. In *Proceedings of IEEE-CVPR*, pages 355–362, 2001.

[92] M. Bertalmío, V. Caselles, S. Masnou, and G. Sapiro. *Encyclopedia of Computer Vision*, chapter Inpainting. Springer, 2014.

[93] M. Bertalmío, V. Caselles, and A. Pardo. Movie denoising by average of warped lines. *IEEE Transactions on Image Processing*, 16(9):2333–2347, 2007.

[94] M. Bertalmío, V. Caselles, and E. Provenzi. Issues about retinex theory and contrast enhancement. *International Journal of Computer Vision*, 83(1):101–119, 2009.

[95] M. Bertalmío, V. Caselles, E. Provenzi, and A. Rizzi. Perceptual color correction through variational techniques. *IEEE Transactions on Image Processing*, 16:1058–1072, 2007.

[96] M. Bertalmío and J. Cowan. Implementing the Retinex algorithm with Wilson-Cowan equations. *Journal of Physiology, Paris*, 2009.

[97] M. Bertalmío, P. Fort, and D. Sanchez-Crespo. Real-time, accurate depth of field using anisotropic diffusion and programmable graphics cards. In *Proceedings of 2nd International Symposium on 3D Data Processing, Visualization and Transmission (3DPVT)*, pages 767–773, 2004.

[98] M. Bertalmío and S. Levine. Color matching for stereoscopic cinema. In *Proceedings of the 6th International Conference on Computer Vision/Computer Graphics Collaboration Techniques and Applications*, page 6. ACM, 2013.

[99] M. Bertalmío and S. Levine. A variational approach for the fusion of exposure bracketed pairs. *Image Processing, IEEE Transactions on*, 22(2):712–723, 2013.

[100] M. Bertalmío and S. Levine. Denoising an image by denoising its curvature image. *SIAM SIIMS, to appear*, 2014.

[101] M. Bertalmío, G. Sapiro, V. Caselles, and C. Ballester. Image inpainting. In *Proceedings of the 27th annual conference on Computer graphics and interactive techniques*, pages 417–424. ACM Press/Addison-Wesley Publishing Co., 2000.

[102] M. Bertalmío, L. Vese, G. Sapiro, and S. Osher. Simultaneous structure and texture image inpainting. *IEEE Transactions on Image Processing*, 12(8):882–889, 2003.

[103] S. Bianco, A. Bruna, F. Naccari, and R. Schettini. Color space transformations for digital photography exploiting information about the illuminant estimation process. *Journal of the Optical Society of America A*, 29(3):374–384, 2012.

[104] S. Bianco, F. Gasparini, A. Russo, and R. Schettini. A new method for RGB to XYZ transformation based on pattern search optimization. *IEEE Transactions on Consumer Electronics*, 53(3):1020–1028, 2007.

[105] N. Bonnier, F. Schmitt, H. Brettel, and S. Berche. Evaluation of spatial gamut mapping algorithms. In *Proceedings of the 14th Color Imaging Conference*, 2006.

[106] A. Bowyer. Computing Dirichlet tessellations. *The Computer Journal*, 24(2):162–166, 1981.

[107] D.H. Brainard and W.T. Freeman. Bayesian color constancy. *Journal of the Optical Society of America A*, 14(7):1393–1411, 1997.

[108] P.C. Bressloff, J.D. Cowan, M. Golubitsky, P.J. Thomas, and M.C. Wiener. What Geometric Visual Hallucinations Tell Us about the Visual Cortex. *Neural Computation*, 14(3):473–491, 2002.

[109] A. Buades, B. Coll, and J.M. Morel. A non-local algorithm for image denoising. In *IEEE Computer Society Conference on Computer Vision and Pattern Recognition (CVPR)*, volume 2, pages 60–65. IEEE, 2005.

[110] A. Buades, B. Coll, and J.M. Morel. A review of image denoising algorithms, with a new one. *SIAM Multiscale Modeling and Simulation*, 4(2):490–530, 2005.

[111] A. Buades, B. Coll, J.M. Morel, and C. Sbert. Self-similarity driven color demosaicking. *IEEE Transactions on Image Processing*, 18(6):1192–1202, 2009.

[112] G. Buchsbaum. A spatial processor model for object colour perception. *Journal of the Franklin Institute*, 310:337–350, 1980.

[113] A. Bugeau, M. Bertalmío, V. Caselles, and G. Sapiro. A comprehensive framework for image inpainting. *IEEE Transactions on Image Processing*, 19(10):2634–2645, 2010.

[114] V.C. Cardei and B. Funt. Committee-based color constancy. In *IS&T/SIDs Color Imaging Conference*, pages 311–313, 1999.

[115] CCITT. *ITU -Recommendation T.81*, 1993.

[116] S.H. Chan, R. Khoshabeh, K.B. Gibson, P.E. Gill, and T.Q. Nguyen. An augmented Lagrangian method for total variation video restoration. *IEEE Transactions on Image Processing*, 20(11):3097–3111, 2011.

[117] D.M. Chandler. Seven challenges in image quality assessment: Past, present, and future research. *ISRN Signal Processing*, 2013, 2013.

[118] P. Chatterjee and P. Milanfar. Is denoising dead? *IEEE Transactions on Image Processing*, 19(4):895–911, 2010.

[119] B.N. Chen and Y. Lin. Temporal error concealment using selective motion field interpolation. *Electronics Letters*, 42(24), Nov. 2006.

[120] C.M. Cheng, S.J. Lin, and S.H. Lai. Spatio-temporally consistent novel view synthesis algorithm from video-plus-depth sequences for autostereoscopic displays. *IEEE Transactions on Broadcasting*, 57(2):523–532, 2011.

[121] V. Cheung, B.J. Frey, and N. Jojic. Video epitomes. In *IEEE Conference on Computer Vision and Pattern Recognition (CVPR)*, volume 1, pages 42–49, 2005.

[122] K. Chiu, M. Herf, P. Shirley, S. Swamy, C. Wang, and K. Zimmerman. Spatially nonuniform scaling functions for high contrast images. In *Proceedings of Graphics Interface 93*, pages 245–253. Morgan Kaufmann, 1993.

[123] S. Cho, J. Wang, and S. Lee. Video deblurring for hand-held cameras using patch-based synthesis. *ACM Transactions on Graphics (TOG)*, 31(4):64, 2012.

[124] W.H. Cho and K.S. Hong. Affine motion based CMOS distortion analysis and CMOS digital image stabilization. *IEEE Transactions on Consumer Electronics*, 53(3):833–841, 2007.

[125] Y.H. Cho, H.Y. Lee, and D.S. Park. Temporal frame interpolation based on multi-frame feature trajectory. *IEEE Transactions on Circuits and Systems for Video Technology*, 2013.

[126] S. Choi, B. Ham, and K. Sohn. Space-time hole filling with random walks in view extrapolation for 3d video. *IEEE Transactions on Image Processing*, 22(6):2429–2441, 2013.

[127] T.J. Cholewo and S. Love. Gamut boundary determination using alpha-shapes. In *Proceedings of IS&T/SID Seventh Color Imaging Conference: Color Science, Systems and Applications*, pages 200–204, 1999.

[128] K.H. Chung and Y.H. Chan. Color demosaicing using variance of color differences. *IEEE Trans. Image Processing*, pages 2944–55, 2006.

[129] A. Ciomaga, P. Monasse, and J-M. Morel. Level lines shortening yields an image curvature microscope. In *2010 17th IEEE International Conference on Image Processing (ICIP)*, pages 4129–4132. IEEE, 2010.

[130] J. Coghill. Digital imaging technology 101. Technical report, DALSA, 2003.

[131] M. Cowan, G. Kennel, T. Maier, and B. Walker. Contrast sensitivity experiment to determine the bit depth for digital cinema. *SMPTE motion imaging journal*, 113(9):281–292, 2004.

[132] I.J. Cox, S.L. Hingorani, S.B. Rao, and B.M. Maggs. A maximum likelihood stereo algorithm. *Computer Vision and Image Understanding*, 63(3):542–567, 1996.

[133] A. Criminisi, P. Pérez, and K. Toyama. Region filling and object removal by exemplar-based inpainting. *IEEE Transactions on Image Processing*, 13(9):1200–1212, 2004.

[134] B. Crowell. *Optics*, volume 5. Light and Matter, 2000.

[135] S. Cvetkovic, H. Jellema, et al. Automatic level control for video cameras towards HDR techniques. *Journal on Image and Video Processing*, 2010:13, 2010.

[136] P. Cyriac, T. Batard, and M. Bertalmío. A variational method for the optimization of tone mapping operators. In *Proceedings of PSIVT 2013*, Lecture Notes in Computer Science. Springer.

[137] K. Dabov, A. Foi, V. Katkovnik, and K. Egiazarian. Image denoising by sparse 3D transform-domain collaborative filtering. *Image Processing, IEEE Transactions on*, 16(8):2080–2095, 2007.

[138] S. Daly. *Digital images and human vision*, chapter The visible differences predictor: an algorithm for the assessment of image fidelity, pages 179–206. MIT Press, Cambridge, MA, USA, 1993.

[139] A. Danielyan, A. Foi, V. Katkovnik, and K. Egiazarian. Image and video super-resolution via spatially adaptive block-matching filtering. In *Proceedings of International Workshop on Local and Non-Local Approximation in Image Processing, Switzerland*, 2008.

[140] G. De Haan and E.B. Bellers. De-interlacing of video data. *IEEE Transactions on Consumer Electronics*, 43:819 – 825, August 1997.

[141] G. De Haan, P. Biezen, H. Huijgen, and O.A. Ojo. True-motion estimation with 3D recursive search block matching. *IEEE Transactions on Circuits and Systems for Video Technology*, 3(5):368–379, 1993.

[142] P. Debevec and J. Malik. Recovering high dynamic range radiance maps from photographs. In *Proceedings of the 24th annual conf. on Computer graphics*, pages 369–378, 1997.

[143] S. Devaud. *Tourner en vidéo HD avec les reflex Canon: EOS 5D Mark II, EOS 7D, EOS 1D Mark IV*. Eyrolles, 2010.

[144] S. Dikbas and Y. Altunbasak. Novel true-motion estimation algorithm and its application to motion-compensated temporal frame interpolation. *IEEE Transactions on Image Processing*, 22(8):2931–2945, 2013.

[145] D.L. Donoho. De-noising by soft-thresholding. *IEEE Transactions on Information Theory*, 41(3):613–627, 1995.

[146] F. Drago, K. Myszkowski, T. Annen, and N. Chiba. Adaptive logarithmic mapping for displaying high contrast scenes. *Computer Graphics Forum*, 22:419–426, 2003.

[147] F. Dugay, I. Farup, and J.Y. Hardeberg. Perceptual evaluation of color gamut mapping algorithms. *Color Research & Application*, 33(6):470–476, 2008.

[148] F. Durand and J. Dorsey. Fast bilateral filtering for the display of high-dynamic-range images. pages 257–266. SIGGRAPH 2002, Proceedings of the 29th annual conference on Computer Graphics and interactive techniques, 2002.

[149] M. Ebdelli, C. Guillemot, and O. Le Meur. Examplar-based video inpainting with motion-compensated neighbor embedding. In *19th IEEE International Conference on Image Processing (ICIP)*, pages 1737–1740. IEEE, 2012.

[150] M. Ebner. *Color constancy.* Wiley, 2007.

[151] F. Ebrahimi, M. Chamik, and S. Winkler. JPEG vs. JPEG 2000: an objective comparison of image encoding quality. In *Proceedings of SPIE*, volume 5558, pages 300–308, 2004.

[152] M. Ebrahimi and E.R. Vrscay. Multi-frame super-resolution with no explicit motion estimation. In *Proceedings of the 2008 International Conference on Image Processing, Computer Vision, and Pattern Recognition (IPCV2008)*, pages 455–459, 2008.

[153] A. Efros and T. Leung. Texture synthesis by non-parametric sampling. In *IEEE International Conference on Computer Vision*, pages 1033–1038, 1999.

[154] M. Elad and M. Aharon. Image denoising via sparse and redundant representations over learned dictionaries. *IEEE Transactions on Image Processing*, 15(12):3736–3745, 2006.

[155] M.D. Fairchild. Color appearance models: CIECAM02 and beyond. In *tutorial notes, IS&T/SID 12th Color Imaging Conference*, 2004.

[156] M.D. Fairchild. *Color appearance models.* J. Wiley, 2005.

[157] M.D. Fairchild and G.M. Johnson. The iCAM framework for image appearance, image differences, and image quality. *Journal of Electronic Imaging*, 13(1):126–138, 2004.

[158] H.S. Fairman, M.H. Brill, and H. Hemmendinger. How the CIE 1931 color-matching functions were derived from Wright-Guild data. *Color Research and Application*, 22(1):11–23, 1997.

[159] Z. Farbman and D. Lischinski. Tonal stabilization of video. *ACM Transactions on Graphics (TOG)*, 30(4):89, 2011.

[160] I. Farup, C. Gatta, and A. Rizzi. A multiscale framework for spatial gamut mapping. *IEEE Transactions on Image Processing*, 16(10):2423–2435, 2007.

[161] R. Fattal, D. Lischinski, and M. Werman. Gradient domain high dynamic range compression. In *ACM Transactions Graphics*, volume 21 (3), pages 249–256, 2002.

[162] O. Faugeras, T. Viéville, E. Theron, J. Vuillemin, B. Hotz, Z. Zhang, L. Moll, P. Bertin, H. Mathieu, P. Fua, et al. Real-time correlation-based stereo: algorithm, implementations and applications. *INRIA Technical Report*, 1993.

[163] S. Ferradans, M. Bertalmío, and V. Caselles. Geometry-based demosaicking. *IEEE Transactions on Image Processing*, 18(3):665–670, 2009.

[164] S. Ferradans, M. Bertalmío, E. Provenzi, and V. Caselles. An analysis of visual adaptation and contrast perception for tone mapping. *IEEE Transactions on Pattern Analysis and Machine Intelligence*, October 2011.

[165] S. Ferradans, M. Bertalmío, E. Provenzi, and V. Caselles. Generation of HDR images in non-static conditions based on gradient fusion. In *International Conference on Computer Vision Theory and Applications (VISAPP)*, pages 31–37, 2012.

[166] G. Finlayson and M.S. Drew. Constrained least-squares regression in color spaces. *Journal of Electronic Imaging*, 6(4):484–493, 1997.

[167] G. Finlayson, M.S. Drew, and B.V. Funt. Diagonal transforms suffice for color constancy. In *Proceedings of Fourth International Conference on Computer Vision*, pages 164–171. IEEE, 1993.

[168] G. Finlayson, M.S. Drew, and B.V. Funt. Spectral sharpening: sensor transformations for improved color constancy. *Journal of the Optical Society of America A*, 11(5):1553–1563, 1994.

[169] G. Finlayson and S. Hordley. Improving gamut mapping color constancy. *IEEE Transactions on Image Processing*, 9(10):1774–1783, 2000.

[170] G. Finlayson, S.D. Hordley, and P.M. Hubel. Color by correlation: A simple, unifying framework for color constancy. *IEEE Transactions on Pattern Analysis and Machine Intelligence*, 23(11):1209–1221, 2001.

[171] G. Finlayson and G. Schaefer. Solving for colour constancy using a constrained dichromatic reflection model. *International Journal of Computer Vision*, 42(3):127–144, 2001.

[172] G. Finlayson and E. Trezzi. Shades of gray and colour constancy. In *Twelfth Color Imaging Conference: Color Science and Engineering Systems, Technologies, and Applications, Scottsdale, Arizona.* University of East Anglia, 2004.

[173] P. Forssen and E. Ringaby. Rectifying rolling shutter video from hand-held devices. In *2010 IEEE Conference on Computer Vision and Pattern Recognition (CVPR)*, pages 507–514. IEEE, 2010.

[174] D.A. Forsyth. A novel algorithm for color constancy. *International Journal of Computer Vision*, 5(1):5–35, 1990.

[175] D.H. Foster. Color constancy. *Vision research*, 51(7):674–700, 2011.

[176] G. Freedman and R. Fattal. Image and video upscaling from local self-examples. *ACM Transactions on Graphics (TOG)*, 30(2):12, 2011.

[177] W.T. Freeman, T.R. Jones, and E.C. Pasztor. Example-based super-resolution. *Computer Graphics and Applications, IEEE*, 22(2):56–65, 2002.

[178] J.P. Frisby and J.V. Stone. *Seeing: the computational approach to biological vision*. The MIT Press, 2010.

[179] D. Gao, X. Wu, G. Shi, and L. Zhang. Color demosaicking with an image formation model and adaptive pca. *Journal of Visual Communication and Image Representation*, 2012.

[180] B. Gardner. Perception and the art of 3D storytelling. *Creative COW Magazine*, May/June 2009.

[181] M. Ghodstinat, A. Bruhn, and J. Weickert. Deinterlacing with motion-compensated anisotropic diffusion. In *Statistical and Geometrical Approaches to Visual Motion Analysis*, pages 91–106. Springer, 2009.

[182] A. Gijsenij, T. Gevers, and J. Van De Weijer. Computational color constancy: Survey and experiments. *IEEE Transactions on Image Processing*, 20(9):2475–2489, 2011.

[183] A.L. Gilchrist. *Lightness, Brightness and Transparency.* Psychology Press, 1994.

[184] E.J. Giorgianni, T.E. Madden, and K.E. Spaulding. Color management for digital imaging systems. *Digital color imaging handbook*, 2003.

[185] A. Goldstein and R. Fattal. Video stabilization using epipolar geometry. *ACM Transactions on Graphics (TOG)*, 31(5):126, 2012.

[186] W. Gompertz. *What are You Looking At?: 150 Years of Modern Art in the Blink of an Eye.* Penguin UK, 2012.

[187] M. Granados, K.I. Kim, J. Tompkin, J. Kautz, and C. Theobalt. Background inpainting for videos with dynamic objects and a free-moving camera. In *Proceedings of the 2012 European Conference on Computer Vision (ECCV)*, pages 682–695. Springer, 2012.

[188] M. Grundmann, V. Kwatra, D. Castro, and I. Essa. Calibration-free rolling shutter removal. In *Computational Photography (ICCP), 2012 IEEE International Conference on*, pages 1–8. IEEE, 2012.

[189] M. Grundmann, V. Kwatra, and I. Essa. Auto-directed video stabilization with robust l1 optimal camera paths. In *IEEE Conference on Computer Vision and Pattern Recognition (CVPR)*, pages 225–232. IEEE, 2011.

[190] Y. HaCohen, E. Shechtman, D.B. Goldman, and D. Lischinski. Non-rigid dense correspondence with applications for image enhancement. *ACM Transactions on Graphics (TOG)*, 30(4):70, 2011.

[191] J. Hahn, X.C. Tai, S. Borok, and A.M. Bruckstein. Orientation-matching minimization for image denoising and inpainting. *International Journal of Computer Vision*, 92(3):308–324, 2011.

[192] J.Y. Hardeberg, E. Bando, and M. Pedersen. Evaluating colour image difference metrics for gamut-mapped images. *Coloration Technology*, 124(4):243–253, 2008.

[193] G. Haro, M. Bertalmío, and V. Caselles. Visual acuity in day for night. *International Journal of Computer Vision*, 69(1):109–117, 2006.

[194] R. Hartley and A. Zisserman. *Multiple view geometry in computer vision*, volume 2. Cambridge Univ Press, 2000.

[195] A. Hervieu, N. Papadakis, A. Bugeau, P. Gargallo, and V. Caselles. Stereoscopic image inpainting: distinct depth maps and images inpainting. In *20th International Conference on Pattern Recognition (ICPR)*, pages 4101–4104. IEEE, 2010.

[196] K. Hirakawa and T.W. Parks. Adaptive homogeneity-directed demosaicing algorithm. *IEEE Trans. Image Processing*, 14(3):360–369, 2005.

[197] K. Hirakawa and T.W. Parks. Joint demosaicing and denoising. *IEEE Transactions on Image Processing*, 15(8):2146–2157, 2006.

[198] M. Hirsch, S. Sra, B. Scholkopf, and S. Harmeling. Efficient filter flow for space-variant multiframe blind deconvolution. In *IEEE Conference on Computer Vision and Pattern Recognition (CVPR)*, pages 607–614. IEEE, 2010.

[199] S.D. Hordley. Scene illuminant estimation: past, present, and future. *Color Research & Application*, 31(4):303–314, 2006.

[200] J. Hu, O. Gallo, K. Pulli, and X. Sun. HDR deghosting: How to deal with saturation? In *Proceedings of the IEEE Conference on Computer Vision and Pattern Recognition (CVPR)*. IEEE, June 2013.

[201] A.M. Huang and T. Nguyen. Correlation-based motion vector processing with adaptive interpolation scheme for motion-compensated frame interpolation. *IEEE Transactions on Image Processing*, 18(4):740–752, 2009.

[202] T.W. Huang and H.T. Chen. Landmark-based sparse color representations for color transfer. In *Computer Vision, 2009 IEEE 12th International Conference on*, pages 199–204. IEEE, 2009.

[203] D.H. Hubel. *Eye, Brain, and Vision.* Scientific American Library, 1995.

[204] P.M. Hubel, J. Holm, G.D. Finlayson, M.S. Drew, et al. Matrix calculations for digital photography. In *The Fifth Color Imaging Conference: Color, Science, Systems and Applications*, 1997.

[205] P. Hung. *Image Sensors and Signal Processing for Digital Still Cameras*, chapter Color theory and its application to digital still cameras, pages 205–221. CRC, 2005.

[206] M. Irani. Multi-frame correspondence estimation using subspace constraints. *International Journal of Computer Vision*, 48(3):173–194, 2002.

[207] N. Jacobson, Y.L. Lee, V. Mahadevan, N. Vasconcelos, and T.Q. Nguyen. A novel approach to FRUC using discriminant saliency and frame segmentation. *IEEE Transactions on Image Processing*, 19(11):2924–2934, 2010.

[208] J. Janesick. Dueling detectors. CMOS or CCD? *SPIE OE magazine*, 41:30–33, 2002.

[209] J. Jeon, J. Lee, and J. Paik. Robust focus measure for unsupervised auto-focusing based on optimum discrete cosine transform coefficients. *IEEE Transactions on Consumer Electronics*, 57(1):1–5, 2011.

[210] J. Jia and C.K. Kang. Inference of segmented color and texture description by tensor voting. *IEEE Transactions on Pattern Analysis and Machine Intelligence*, 26(6):771–786, 2004.

[211] J. Jia, Y.W. Tai, T.P. Wu, and C.K. Tang. Video repairing under variable illumination using cyclic motions. *IEEE Transactions on Pattern Analysis and Machine Intelligence*, 28(5):832–83, May 2006.

[212] J. Jia, T. Wu, Y. Tai, and C. Tang. Video repairing under variable illumination using cyclic motions. *Proceedings of 2004 IEEE Computer Society Conference on Computer Vision and Pattern Recognition*, 1:364–371, 2004.

[213] D.B. Judd. Hue, saturation, and lightness of surface colors with chromatic illumination. *Journal of the Optical Society of America*, 30(1):2–32, 1940.

[214] D.B. Judd. *Contributions to color science*, volume 545, chapter Color appearance, pages 539–564. National Bureau of Standards (USA), 1979.

[215] D.B. Judd. *Contributions to color science*, volume 545, chapter Estimation of chromaticity differences and nearest color temperature on the standard 1931 ICI colorimetric coordinate system, pages 207–212. National Bureau of Standards (USA), 1979.

[216] D.B. Judd. *Contributions to color science*, volume 545, chapter The unsolved problem of color perception, pages 516–522. National Bureau of Standards (USA), 1979.

[217] D.B. Judd. *Contributions to color science*, volume 545, chapter Appraisal of Land's work on two-primary color projections, pages 471–486. National Bureau of Standards (USA), 1979.

[218] D.B. Judd. *Contributions to color science*, volume 545. National Bureau of Standards (USA), 1979.

[219] S. Kagarlitsky, Y. Moses, and Y. Hel-Or. Piecewise-consistent color mappings of images acquired under various conditions. In *IEEE 12th International Conference on Computer Vision*, pages 2311–2318. IEEE, 2009.

[220] B.H. Kang, J. Morovič, M.R. Luo, and M.S. Cho. Gamut compression and extension algorithms based on observer experimental data. *ETRI journal*, 25(3):156, 2003.

[221] S.B. Kang, M. Uyttendaele, S. Winder, and R. Szeliski. High dynamic range video. *ACM Transactions on Graphics*, 22(3):319–325, 2003.

[222] S.J. Kang, S. Yoo, and Y.H. Kim. Dual motion estimation for frame rate up-conversion. *IEEE Transactions on Circuits and Systems for Video Technology*, 20(12):1909–1914, 2010.

[223] W.C. Kao, S.H. Wang, L.Y. Chen, and S.Y. Lin. Design considerations of color image processing pipeline for digital cameras. *IEEE Transactions on Consumer Electronics*, 52(4):1144–1152, 2006.

[224] V. Katkovnik, A. Danielyan, and K. Egiazarian. Decoupled inverse and denoising for image deblurring: variational BM3D-frame technique. In *18th IEEE International Conference on Image Processing (ICIP)*, pages 3453–3456. IEEE, 2011.

[225] S.H. Keller, F. Lauze, and M. Nielsen. Deinterlacing using variational methods. *IEEE Transactions on Image Processing*, 17(11):2015–2028, 2008.

[226] G. Kennel. *Color and Mastering for Digital Cinema (Digital Cinema Industry Handbook Series).* Taylor and Francis US, 2006.

[227] M.C. Kim, Y.C. Shin, Y.R. Song, S.J. Lee, and I.D. Kim. Wide gamut multi-primary display for HDTV. In *European Conference on Colour in Graphics, Imaging, and Vision*, pages 248–253, 2004.

[228] S. Kim, H. Lin, Z. Lu, S. Susstrunk, S. Lin, and M. Brown. A new in-camera imaging model for color computer vision and its application. *IEEE Transactions on Pattern Analysis and Machine Intelligence*, 2012.

[229] S. Kim, Y.W. Tai, S.J. Kim, M.S. Brown, and Y. Matsushita. Nonlinear camera response functions and image deblurring. In *IEEE Conference on Computer Vision and Pattern Recognition (CVPR)*, pages 25–32. IEEE, 2012.

[230] R. Kimmel, D. Shaked, M. Elad, and I. Sobel. Space-dependent color gamut mapping: a variational approach. *IEEE Transactions on Image Processing*, 14(6):796–803, 2005.

[231] Kodak. `http://r0k.us/graphics/kodak/`.

[232] A.C. Kokaram. *Motion picture restoration: digital algorithms for artefact suppression in degraded motion picture film and video.* Springer-Verlag London, UK, 1998.

[233] S.J. Koppal, C.L. Zitnick, M. Cohen, S.B. Kang, B. Ressler, and A. Colburn. A viewer-centric editor for 3D movies. *Computer Graphics and Applications, IEEE*, 31(1):20–35, 2011.

[234] H. Kotera. A scene-referred color transfer for pleasant imaging on display. In *Proceedings of IEEE International Conference on Image Processing (ICIP)*, pages 5–8, 2005.

[235] T. Koyama. *Image sensors and signal processing for digital still cameras*, chapter Optics in Digital Still Cameras. CRC, 2005.

[236] W. Kress and M. Stevens. Derivation of 3-dimensional gamut descriptors for graphic arts output devices. In *Proceedings of TAGA*, pages 199–214. Technical association of the graphic arts, 1994.

[237] J. Kuang, G.M. Johnson, and M.D. Fairchild. iCAM06: A refined image appearance model for HDR image rendering. *Journal of Visual Communication and Image Representation*, 18(5):406–414, 2007.

[238] J. Laird, R. Muijs, and J. Kuang. Development and evaluation of gamut extension algorithms. *Color Research & Application*, 34(6):443–451, 2009.

[239] E.Y. Lam and G.S.K. Fung. *Single-sensor imaging: Methods and applications for digital cameras*, chapter Automatic white balancing in digital photography, pages 267–294. CRC Press, 2008.

[240] W.M. Lam, A.R. Reibman, and B. Liu. Recovery of lost or erroneously received motion vectors. In *Proceedings of IEEE International Conference on Acoustics, Speech, and Signal Processing (ICASSP)*, 1993.

[241] M. Lambooij, W. IJsselsteijn, I. Heynderickx, et al. Visual discomfort in stereoscopic displays: a review. *Stereoscopic Displays and Virtual Reality Systems XIV*, 6490(1), 2007.

[242] E.H. Land. Some comments on Dr. Judd's Paper. *Journal of the Optical Society of America*, 50(3):268, 1960.

[243] E.H. Land. The Retinex theory of color vision. *Scientific American*, 237:108–128, 1977.

[244] E.H. Land and J.J. McCann. Lightness and Retinex theory. *Journal of the Optical Society of America*, 61(1):1–11, January 1971.

[245] C. Lau, W. Heidrich, and R. Mantiuk. Cluster-based color space optimizations. In *Proceedings of IEEE International Conference on Computer Vision*, ICCV '11, pages 1172–1179, 2011.

[246] M. Lebrun, M. Colom, A. Buades, and J.M. Morel. Secrets of image denoising cuisine. *Acta Numerica*, 21(1):475–576, 2012.

[247] K. Lee and C. Lee. High quality spatially registered vertical temporal filtering for deinterlacing. *IEEE Transactions on Consumer Electronics*, 59(1):182–190, 2013.

[248] K.Y. Lee, Y.Y. Chuang, B.Y. Chen, and M. Ouhyoung. Video stabilization using robust feature trajectories. In *IEEE 12th International Conference on Computer Vision*, pages 1397–1404. IEEE, 2009.

[249] A. Levin and B. Nadler. Natural image denoising: Optimality and inherent bounds. In *2011 IEEE Conference on Computer Vision and Pattern Recognition (CVPR)*, pages 2833–2840. IEEE, 2011.

[250] X. Li and Y. Zheng. Patch-based video processing: a variational bayesian approach. *Circuits and Systems for Video Technology, IEEE Transactions on*, 19(1):27–40, 2009.

[251] N.X. Lian, L. Chang, Y.P. Tan, and V. Zagorodnov. Adaptive filtering for color filter array demosaicking. *IEEE Transactions on Image Processing*, 16(10):2515–2525, Oct. 2007.

[252] C.K. Liang, L.W. Chang, and H.H. Chen. Analysis and compensation of rolling shutter effect. *IEEE Transactions on Image Processing*, 17(8):1323–1330, 2008.

[253] D. Lischinski, Z. Farbman, M. Uyttendaele, and R. Szeliski. Interactive local adjustment of tonal values. In *SIGGRAPH '06: ACM SIGGRAPH 2006 Papers*, pages 646–653, New York, NY, USA, 2006. ACM.

[254] I. Lissner, J. Preiss, P. Urban, M. S. Lichtenauer, and P. Zolliker. Image-difference prediction: From grayscale to color. *IEEE Transactions on Image Processing*, 22(2):435–446, 2013.

[255] F. Liu, M. Gleicher, H. Jin, and A. Agarwala. Content-preserving warps for 3D video stabilization. In *ACM Transactions on Graphics (TOG)*, volume 28, page 44. ACM, 2009.

[256] F. Liu, M. Gleicher, J. Wang, H. Jin, and A. Agarwala. Subspace video stabilization. *ACM Transactions on Graphics (TOG)*, 30(1):4, 2011.

[257] H. Liu, R. Xiong, D. Zhao, S. Ma, and W. Gao. Multiple hypotheses bayesian frame rate up-conversion by adaptive fusion of motion-compensated interpolations. *IEEE Transactions on Circuits and Systems for Video Technology*, 22(8):1188–1198, 2012.

[258] D.G. Lowe. Distinctive image features from scale-invariant keypoints. *International journal of computer vision*, 60(2):91–110, 2004.

[259] S. Lumet. *Making movies.* Alfred A. Knopf, 1995.

[260] M. Lysaker, S. Osher, and X.C. Tai. Noise removal using smoothed normals and surface fitting. *Image Processing, IEEE Transactions on*, 13(10):1345–1357, 2004.

[261] J. Mairal, M. Elad, and G. Sapiro. Sparse representation for color image restoration. *IEEE Transactions on Image Processing*, 17(1):53–69, 2008.

[262] R. Mantiuk, S.J. Daly, K. Myszkowski, and H.P. Seidel. Predicting visible differences in high dynamic range images: model and its calibration. In *Proceedings SPIE*, volume 5666, pages 204–214, 2005.

[263] R. Mantiuk, K. Myszkowski, and H.P. Seidel. A perceptual framework for contrast processing of high dynamic range images. *ACM Transactions on Applied Perception (TAP)*, 3 (3):286–308, 2006.

[264] Rafał Mantiuk, Scott Daly, and Louis Kerofsky. Display adaptive tone mapping. In *ACM Transactions on Graphics (TOG)*, volume 27, page 68. ACM, 2008.

[265] S. Masnou and J. Morel. Level-lines based disocclusion. *IEEE International Conference on Image Processing*, 1998.

[266] Y. Matsushita, E. Ofek, X. Tang, and H.Y. Shum. Full-frame video stabilization. In *Proceedings of the IEEE Computer Society Conference on Computer Vision and Pattern Recognition (CVPR'05)*, volume 1, pages 50 – 57, June 2005.

[267] J. McCann. *Colour image science: exploiting digital media*, chapter A spatial colour gamut calculation to optimise colour appearance. Wiley, 2002.

[268] J. McCann, C. Parraman, and A. Rizzi. Reflectance, illumination and edges. *Proceedings Color and Imaging Conference (CIC)*, 201(17):2–7, 2009.

[269] J.J. McCann and A. Rizzi. *The art and science of HDR imaging.* Wiley, 2012.

[270] B. Mendiburu. *3D movie making: stereoscopic digital cinema from script to screen.* Focal Press, 2009.

[271] D. Menon, S. Andriani, and G. Calvagno. Demosaicing With Directional Filtering and a posteriori Decision. *IEEE Transactions on Image Processing*, 16(1):132–141, 2007.

[272] D. Menon and G. Calvagno. Joint demosaicking and denoising with space-varying filters. In *16th IEEE International Conference on Image Processing (ICIP)*, pages 477–480. IEEE, 2009.

[273] D. Menon and G. Calvagno. Color image demosaicking: an overview. *Signal Processing: Image Communication*, 26(8):518–533, 2011.

[274] J. Meyer and B. Barth. Color gamut matching for hard copy. In *SID Digest, Applied Vision Topical Meeting*, pages 86–89, 1989.

[275] T.M. Moldovan, S. Roth, and M.J. Black. Denoising archival films using a learned Bayesian model. In *IEEE International Conference on Image Processing*, pages 2641–2644. IEEE, 2006.

[276] J.D. Mollon. *The science of color*, chapter The origins of modern color science, pages 1–39. Optical Society of America Oxford, 2003.

[277] P. Monnier. Standard definitions of chromatic induction fail to describe induction with s-cone patterned backgrounds. *Vision Research*, 48(27):2708 – 2714, 2008.

[278] E.D. Montag and M.D. Fairchild. Gamut mapping: Evaluation of chroma clipping techniques for three destination gamuts. In *IS&T/SID Sixth Colour Imaging Conference*, pages 57–61, 1998.

[279] C. Morimoto and R. Chellappa. Evaluation of image stabilization algorithms. In *Proceedings of the 1998 IEEE International Conference on Acoustics, Speech and Signal Processing (ICASSP)*, volume 5, pages 2789–2792. IEEE, 1998.

[280] A. Morimura, K. Uomori, Y. Kitamura, A. Fujioka, J. Harada, S. Iwamura, and M. Hirota. A digital video camera system. *IEEE Transactions on Consumer Electronics*, 36(4):3866–3876, 1990.

[281] N. Moroney, M.D. Fairchild, R. Hunt, C. Li, M.R. Luo, and T. Newman. The CIECAM02 color appearance model. In *Proceedings of Color and Imaging Conference*. Society for Imaging Science and Technology, 2002.

[282] J. Morovič. *Color gamut mapping*. John Wiley & Sons, 2008.

[283] J. Morovič and M.R. Luo. Calculating medium and image gamut boundaries for gamut mapping. *Color research and application*, 25(6):394–401, 2000.

[284] J. Morovič and M.R. Luo. The fundamentals of gamut mapping: A survey. *Journal of Imaging Science and Technology*, 45(3):283–290, 2001.

[285] J. Morovič and P. Morovič. Determining colour gamuts of digital cameras and scanners. *Color Research & Application*, 28(1):59–68, 2003.

[286] J. Morovič and Y. Wang. A multi-resolution, full-colour spatial gamut mapping algorithm. In *Proceedings of IST and SIDs 11th Color Imaging Conference*, pages 282–287, 2003.

[287] B. Morse, J. Howard, S. Cohen, and B. Price. Patchmatch-based content completion of stereo image pairs. In *Second International Conference on 3D Imaging, Modeling, Processing, Visualization and Transmission (3DIMPVT)*, pages 555–562. IEEE, 2012.

[288] A. Mosleh, N. Bouguila, and A. Ben Hamza. Video completion using bandlet transform. *IEEE Transactions on Multimedia*, 14(6):1591–1601, 2012.

[289] G. M. Murch and J. M. Taylor. Color in computer graphics: Manipulating and matching color. *Eurographics Seminar: Advances in Computer Graphics V*, pages 41–47, 1989.

[290] K. Myszkowski, R. Mantiuk, and G. Krawczyk. High dynamic range video. *Synthesis Lectures on Computer Graphics and Animation*, 1(1):1–158, 2008.

[291] J. Nakamura. *Image sensors and signal processing for digital still cameras [Chapter 3]*. CRC, 2005.

[292] J. Nakamura. *Image sensors and signal processing for digital still cameras [Chapter 4]*. CRC, 2005.

[293] S. Nakauchi, S. Hatanaka, and S. Usui. Color gamut mapping based on a perceptual image difference measure. *Color Research & Application*, 24(4):280–291, 1999.

[294] P. Ndjiki-Nya, M. Koppel, D. Doshkov, H. Lakshman, P. Merkle, K. Muller, and T. Wiegand. Depth image-based rendering with advanced texture synthesis for 3D video. *IEEE Transactions on Multimedia*, 13(3):453–465, 2011.

[295] H.S. Oh, Y. Kim, Y. Jung, A. W. Morales, and S.J. Ko. Spatio-temporal edge-based median filtering for deinterlacing. In *Digest of Technical Papers of the International Conference on Consumer Electronics (ICCE)*, pages 52 – 53, Jun 2000.

[296] K.J. Oh, S. Yea, and Y.S. Ho. Hole filling method using depth based in-painting for view synthesis in free viewpoint television and 3D video. In *Picture Coding Symposium (PCS)*, pages 1–4. IEEE, 2009.

[297] S. Osher, M. Burger, D. Goldfarb, J. Xu, and W. Yin. An iterative regularization method for total variation-based image restoration. *Multiscale Modeling and Simulation*, 4(2):460, 2005.

[298] O. Packer and D.R. Williams. *The Science of Color*, chapter Light, the retinal image, and photoreceptors, pages 41–102. Elsevier Science Ltd, 2003.

[299] D. Paliy, V. Katkovnik, R. Bilcu, S. Alenius, and K. Egiazarian. Spatially adaptive color filter array interpolation for noiseless and noisy data. *International Journal of Imaging Systems and Technology*, 17(3):105–122, 2007.

[300] R. Palma-Amestoy, E. Provenzi, M. Bertalmío, and V. Caselles. A perceptually inspired variational framework for color enhancement. *IEEE Transactions on Pattern Analysis and Machine Intelligence*, 31(3):458–474, 2009.

[301] N. Paragios, Y. Chen, and O. Faugeras, editors. *Mathematical models in computer vision: the handbook*, chapter PDE-based image and surface inpainting. Springer, 2005.

[302] A. Pardo and G. Sapiro. Visualization of high dynamic range images. *IEEE Transactions on Image Processing*, 12(6):639–647, 2003.

[303] K. Parulski and K. Spaulding. *Digital Color Imaging Handbook*, chapter Color image processing for digital cameras, pages 727–757. Boca Raton, FL: CRC Press, 2003.

[304] S.N. Pattanaik, J. Tumblin, H. Yee, and D.P. Greenberg. Time-dependent visual adaptation for fast realistic image display. In *Proceedings of SIGGRAPH*, pages 47–54, 2000.

[305] K.A. Patwardhan, G. Sapiro, and M. Bertalmío. Video inpainting under constrained camera motion. *IEEE Transactions on Image Processing*, 16(2):545–553, 2007.

[306] M. Pedersen and J.Y. Hardeberg. Full-reference image quality metrics: Classification and evaluation. *Foundations and Trends in Computer Graphics and Vision*, 7(1):1–80, 2011.

[307] P. Pérez, M. Gangnet, and A. Blake. Poisson image editing. In *ACM Transactions on Graphics (TOG)*, volume 22, pages 313–318. ACM, 2003.

[308] P. Perona and J. Malik. Scale-space and edge detection using anisotropic diffusion. *IEEE Transactions on Pattern Analysis and Machine Intelligence*, 12, 1990.

[309] H.A. Peterson, H. Peng, J.H. Morgan, and W.B. Pennebaker. Quantization of color image components in the DCT domain. *Proceedings SPIE 1453, Human Vision, Visual Processing, and Digital Display II*, pages 210–222, 1991.

[310] F. Pitié, A.C. Kokaram, and R. Dahyot. Automated colour grading using colour distribution transfer. *Computer Vision and Image Understanding*, 107(1):123–137, 2007.

[311] F. Porikli. Inter-camera color calibration by correlation model function. In *Proceedings of 2003 International Conference on Image Processing (ICIP)*, volume 2, pages II – 133–6 vol.3, sept. 2003.

[312] J. Portilla, V. Strela, M.J. Wainwright, and E.P. Simoncelli. Image denoising using scale mixtures of Gaussians in the wavelet domain. *IEEE Transactions on Image Processing*, 12(11):1338–1351, 2003.

[313] C. Poynton. *Digital Video and HD: Algorithms and Interfaces.* Morgan Kaufmann, 2003.

[314] C. Poynton. Wide gamut and wild gamut: xvYCC for HD. Poynton's Vector 8 (online), 2010.

[315] M. Protter, M. Elad, H. Takeda, and P. Milanfar. Generalizing the nonlocal-means to super-resolution reconstruction. *Image Processing, IEEE Transactions on*, 18(1):36–51, 2009.

[316] E. Provenzi, L. De Carli, A. Rizzi, and D. Marini. Mathematical definition and analysis of the Retinex algorithm. *Journal of the Optical Society of America A*, 22(12):2613–2621, December 2005.

[317] P.H. Putman. What will replace the CRT for professional video monitors? In *SMPTE Conferences*. Society of Motion Picture and Television Engineers, 2009.

[318] F. Raimbault and A. Kokaram. Stereo-video inpainting. *Journal of Electronic Imaging*, 21(1):011005–1, 2012.

[319] Felix Raimbault, François Pitié, and Anil Kokaram. Stereo video completion for rig and artefact removal. In *13th International Workshop on Image Analysis for Multimedia Interactive Services (WIAMIS)*, pages 1–4. IEEE, 2012.

[320] R. Ramanath, W.E. Snyder, Y. Yoo, and M.S. Drew. Color image processing pipeline. *Signal Processing Magazine, IEEE*, 22(1):34–43, 2005.

[321] S. Rane, G. Sapiro, and M. Bertalmío. Structure and texture filling-in of missing image blocks in wireless transmission and compression applications. *IEEE Trans. Image Processing*, 2003.

[322] S. Reeve and J. Flock. Basic principles of stereoscopic 3D. BSKYB Whitepaper, 2012.

[323] E. Reinhard. Example-based image manipulation. In *6th European Conference on Colour in Graphics, Imaging, and Vision (CGIV 2012)*, Amsterdam, May 2012.

[324] E. Reinhard, M. Adhikhmin, B. Gooch, and P. Shirley. Color transfer between images. *Computer Graphics and Applications, IEEE*, 21(5):34–41, 2001.

[325] E. Reinhard and K. Devlin. Dynamic range reduction inspired by photoreceptor physiology. *IEEE Transactions on Visualization and Computer Graphics*, 11(1):13–24, 2005.

[326] E. Reinhard, M. Stark, P. Shirley, and J. Ferwerda. Photographic tone reproduction for digital images. *ACM Transactions on Graphics*, 21:267–276, 2002.

[327] E. Reinhard, G. Ward, S. Pattanaik, and P. Debevec. *High Dynamic Range Imaging, Acquisition, Display, And Image-Based Lighting.* Morgan Kaufmann Ed., 2005.

[328] A. Rizzi, C. Gatta, and D. Marini. A new algorithm for unsupervised global and local color correction. *Pattern Recognition Letters*, 24:1663–1677, 2003.

[329] S. Roth and M.J. Black. Fields of experts: a framework for learning image priors. In *CVPR 2005*, volume 2 of *IEEE Computer Society Conference on Computer Vision and Pattern Recognition (CVPR)*, pages 860 – 867, 2005.

[330] L.I. Rudin, S. Osher, and E. Fatemi. Nonlinear total variation based noise removal algorithms. *Physica D: Nonlinear Phenomena*, 60(1-4):259–268, 1992.

[331] J. Salvador, A. Kochale, and S. Schweidler. Patch-based spatio-temporal super-resolution for video with non-rigid motion. *Signal Processing: Image Communication*, 2013.

[332] G. Sapiro. Color and illuminant voting. *IEEE Transactions on Pattern Analysis and Machine Intelligence*, 21(11):1210–1215, 1999.

[333] G. Sapiro and V. Caselles. Histogram modification via differential equations. *Journal of Differential Equations*, 135:238–266, 1997.

[334] K. Sato. Image-processing algorithms. *Image sensors and signal processing for digital still cameras, ed. J. Nakamura*, pages 223–253, 2005.

[335] G. Schaefer, S. Hordley, and G. Finlayson. A combined physical and statistical approach to colour constancy. In *IEEE Computer Society Conference on Computer Vision and Pattern Recognition (CVPR)*, volume 1, pages 148–153. IEEE, 2005.

[336] D. Scharstein and R. Szeliski. A taxonomy and evaluation of dense two-frame stereo correspondence algorithms. *International Journal of Computer Vision*, 47(1):7–42, 2002.

[337] C. Schlick. Quantization techniques for visualization of high dynamic range pictures. In *Proceedings of the 5th Eurographics Workshop on Rendering Workshop*, pages 7–20. Springer Verlag, 1994.

[338] M. Schmeing and X. Jiang. Depth image based rendering. In *Pattern Recognition, Machine Intelligence and Biometrics*, pages 279–310. Springer, 2011.

[339] O. Shahar, A. Faktor, and M. Irani. Space-time super-resolution from a single video. In *IEEE Conference on Computer Vision and Pattern Recognition (CVPR)*, pages 3353–3360. IEEE, 2011.

[340] Q. Shan, Z. Li, J. Jia, and C.K. Tang. Fast image/video upsampling. In *ACM Transactions on Graphics (TOG)*, volume 27, page 153. ACM, 2008.

[341] G. Sharma. *Digital Color Imaging Handbook*, chapter Color fundamentals for digital imaging. CRC Press, 2003.

[342] J. Shen. On the foundations of vision modeling: I. Webers law and Weberized TV restoration. *Physica D: Nonlinear Phenomena*, 175(3):241–251, 2003.

[343] Y. Shen, F. Lu, X. Cao, and H. Foroosh. Video completion for perspective camera under constrained motion. In *18th International Conference on Pattern Recognition (ICPR)*, volume 3, pages 63–66. IEEE, 2006.

[344] S.K. Shevell. *The Science of Color*, chapter Color appearance, pages 149–190. Elsevier Science Ltd, 2003.

[345] Y.Q. Shi and H. Sun. *Image and video compression for multimedia engineering: fundamentals, algorithms, and standards, 2nd Edition.* CRC Press, 2008.

[346] T.K. Shih, N.C. Tang, and J.N. Hwang. Exemplar-based video inpainting without ghost shadow artifacts by maintaining temporal continuity. *IEEE Transactions on Circuits and Systems for Video Technology*, 19(3):347–360, 2009.

[347] T. Shiratori, Y. Matsushita, X. Tang, and S.B. Kang. Video completion by motion field transfer. In *2006 IEEE Computer Society Conference on Computer Vision and Pattern Recognition*, volume 1, 2006.

[348] K. Smith, G. Krawczyk, K. Myszkowski, and H.P. Seidel. Beyond tone mapping: Enhanced depiction of tone mapped HDR images. In *Computer Graphics Forum*, pages 427–438, 2006.

[349] S.M. Smith and J.M. Brady. SUSAN - A new approach to low level image processing. *International Journal of Computer Vision*, 23(1):45–78, 1997.

[350] J. L. Starck, M. Elad, and D. L. Donoho. Image decomposition: Separation of texture from piece-wise smooth content. *Proceedings SPIE Annual Meeting*, 2003.

[351] M. Sznaier and O. Camps. Robust identification of periodic systems with applications to texture inpainting. In *44th IEEE Conference on Decision and Control and 2005 European Control Conference (CDC-ECC)*, 2005.

[352] Y.W. Tai, J. Jia, and C.K. Tang. Local color transfer via probabilistic segmentation by expectation-maximization. In *IEEE Computer Society Conference on Computer Vision and Pattern Recognition (CVPR)*, volume 1, pages 747–754. IEEE, 2005.

[353] H. Takeda, S. Farsiu, and P. Milanfar. Deblurring using regularized locally adaptive kernel regression. *IEEE Transactions on Image Processing*, 17(4):550–563, 2008.

[354] H. Takeda, P. Milanfar, M. Protter, and M. Elad. Super-resolution without explicit subpixel motion estimation. *IEEE Transactions on Image Processing*, 18(9):1958–1975, 2009.

[355] D. Tamburrino, D. Alleysson, L. Meylan, and S. Susstrunk. Digital camera workflow for high dynamic range images using a model of retinal processing. In *Proceedings SPIE*, volume 6817, 2008.

[356] N.C. Tang, C.T. Hsu, C.W. Su, T.K. Shih, and H.Y.M. Liao. Video inpainting on digitized vintage films via maintaining spatiotemporal continuity. *IEEE Transactions on Multimedia*, 13(4):602–614, 2011.

[357] CIE Technical Committee. 8 03, Guidelines for the evaluation of gamut mapping algorithms. Technical report, CIE, 2004.

[358] A. Theuwissen. Image sensor architectures for digital cinematography. Technical report, DALSA, 2005.

[359] C. Tomasi and R. Manduchi. Bilateral filtering for gray and color images. In *ICCV '98: Proceedings of the Sixth International Conference on Computer Vision*, pages 839–846, Washington, DC, USA, 1998. IEEE Computer Society.

[360] S. Tominaga and B.A. Wandell. Standard surface-reflectance model and illuminant estimation. *Journal of the Optical Society of America A*, 6(4):576–584, 1989.

[361] C.Y. Tsai and K.T. Song. Heterogeneity-Projection Hard-Decision Color Interpolation Using Spectral-Spatial Correlation. *IEEE Transactions on Image Processing*, 16(1):78–91, 2007.

[362] S. Tsekeridou and I. Patas. MPEG-2 error concealment based on block-matching principles. *IEEE Transactions on Circuits and Systems for Video Technology*, 2000.

[363] J. Tumblin and H. Rushmeier. Tone reproduction for realistic images. *IEEE Computer Graphics and Applications*, pages 42–48, 1993.

[364] J. Tumblin and G. Turk. Lcis: A boundary hierarchy for detail-preserving contrast reduction. In *SIGGRAPH: Conference Proceedings*, pages 83–90, 1999.

[365] B.E. Usevitch. A tutorial on modern lossy wavelet image compression: foundations of JPEG 2000. *Signal Processing Magazine, IEEE*, 18(5):22–35, 2001.

[366] J. Vazquez-Corral, M. Vanrell, R. Baldrich, and F. Tous. Color constancy by category correlation. *IEEE Transactions on Image Processing*, 21(4):1997–2007, 2012.

[367] M.V. Venkatesh, S.C.S Cheung, and J. Zhao. Efficient object-based video inpainting. *Pattern Recognition Letters*, 30(2):168–179, 2009.

[368] C. Wang, L. Zhang, Y. He, and Y.P. Tan. Frame rate up-conversion using trilateral filtering. *IEEE Transactions on Circuits and Systems for Video Technology*, 20(6):886–893, 2010.

[369] D. Wang, A. Vincent, P. Blanchfield, and R. Klepko. Motion-compensated frame rate up-conversion Part II: New algorithms for frame interpolation. *IEEE Transactions on Broadcasting*, 56(2):142–149, 2010.

[370] L. Wang, H. Jin, R. Yang, and M. Gong. Stereoscopic inpainting: Joint color and depth completion from stereo images. In *IEEE Conference on Computer Vision and Pattern Recognition (CVPR)*, pages 1–8. IEEE, 2008.

[371] Y.S. Wang, F. Liu, P.S. Hsu, and T.Y. Lee. Spatially and temporally optimized video stabilization. *IEEE Transactions on Visualization and Computer Graphics*, 19(8):1354–1361, 2013.

[372] Z. Wang and A.C. Bovik. A universal image quality index. *Signal Processing Letters, IEEE*, 9(3):81–84, 2002.

[373] Z. Wang and E.P. Simoncelli. Translation insensitive image similarity in complex wavelet domain. In *Proceedings of IEEE International Conference on Acoustics, Speech, and Signal Processing (ICASSP)*, volume 2, pages 573–576. IEEE, 2005.

[374] G. Ward. *A contrast-based scalefactor for luminance display*, pages 415–421. Academic Press Professional, San Diego, CA, USA, 1994.

[375] G. Ward Larson, H. Rushmeier, and C. Piatko. A visibility matching tone reproduction operator for high dynamic range scenes. *IEEE Transactions on Visualization and Computer Graphics*, 3:291–306, 1997.

[376] WavelengthMedia. CCU Operations. Technical report, http://www.mediacollege.com/video/production/camera-control/, 2012.

[377] C. Weerasinghe, W. Li, I. Kharitonenko, M. Nilsson, and S. Twelves. Novel color processing architecture for digital cameras with CMOS image sensors. *IEEE Transactions on Consumer Electronics*, 51(4):1092–1098, 2005.

[378] Y. Wexler, E. Shechtman, and M. Irani. Space-time video completion. *Proceedings of 2004 IEEE Computer Society Conference on Computer Vision and Pattern Recognition*, 1:120–127, 2004.

[379] Y. Wexler, E. Shechtman, and M. Irani. Space-time completion of video. *IEEE Transactions on Pattern Analysis and Machine Intelligence*, 29(3), March 2007.

[380] H.R. Wilson and J.D. Cowan. Excitatory and inhibitory interactions in localized populations of model neurons. *Biophysical journal*, 12:1–24, 1972.

[381] H.R. Wilson and J.D. Cowan. A mathematical theory of the functional dynamics of cortical and thalamic nervous tissue. *Biological Cybernetics*, 13(2):55–80, 1973.

[382] J.A. Worthey. Limitations of color constancy. *Journal of the Optical Society of America A*, 2(7):1014–1026, 1985.

[383] J.A. Worthey and M.H. Brill. Heuristic analysis of von Kries color constancy. *Journal of the Optical Society of America A*, 3(10):1708–1712, 1986.

[384] W.D. Wright. A re-determination of the trichromatic coefficients of the spectral colours. *Transactions of the Optical Society*, 30(4):141, 1929.

[385] X. Wu and L. Zhang. Improvement of color video demosaicking in temporal domain. *IEEE Transactions on Image Processing*, 15(10):3138–3151, 2006.

[386] G. Wyszecki and W. S. Stiles. *Color science: Concepts and methods, quantitative data and formulas.* John Wiley & Sons, 1982.

[387] Y. Xu, A.J. Mierop, and A.J.P. Theuwissen. Charge domain interlace scan implementation in a CMOS image sensor. *Sensors Journal, IEEE*, 11(11):2621–2627, 2011.

[388] F. Xue, F. Luisier, and T. Blu. Multi-Wiener sure-let deconvolution. *IEEE Transactions on Image Processing*, 22(5):1954–1968, 2013.

[389] L.P. Yaroslavsky. Digital picture processing. An introduction. *Springer Series in Information Sciences, Vol. 9. Springer-Verlag, Berlin-Heidelberg-New York-Tokyo. ISBN 0-387-11934-5 (USA).*, 1, 1985.

[390] L.P. Yaroslavsky. Local adaptive image restoration and enhancement with the use of DFT and DCT in a running window. In *SPIE's 1996 International Symposium on Optical Science, Engineering, and Instrumentation*, pages 2–13. International Society for Optics and Photonics, 1996.

[391] H. Yoo and J. Jeong. Direction-oriented interpolation and its application to de-interlacing. *IEEE Transactions on Consumer Electronics*, 48(4):954 – 962, 2002.

[392] Jia. Y.T., S.M. Hu, and R.R. Martin. Video completion using tracking and fragment merging. In *Pacific Graphics 2005*, volume 21, pages 601–610. The Visual Computer, Springer-Verlag, 2005.

[393] G. Yu, G. Sapiro, and S. Mallat. Solving inverse problems with piecewise linear estimators: from Gaussian mixture models to structured sparsity. *IEEE Transactions on Image Processing*, 21(5):2481–2499, 2012.

[394] S.W. Zamir, J. Vazquez-Corral, and M. Bertalmío. Gamut mapping through perceptually-based contrast reduction. In *Proceedings of PSIVT 2013*, Lecture Notes in Computer Science. Springer, 2013.

[395] J. Zhai, K. Yu, J. Li, and S. Li. A low complexity motion compensated frame interpolation method. In *IEEE International Symposium on Circuits and Systems (ISCAS)*, pages 4927–4930. IEEE, 2005.

[396] G. Zhang, W. Hua, X. Qin, Y. Shao, and H. Bao. Video stabilization based on a 3D perspective camera model. *The Visual Computer*, 25(11):997–1008, 2009.

[397] H. Zhang, J. Yang, Y. Zhang, and T.S. Huang. Non-local kernel regression for image and video restoration. In *Proceedings of the European Conference on Computer Vision (ECCV)*, pages 566–579. Springer, 2010.

[398] L. Zhang, W. Dong, X. Wu, and G. Shi. Spatial-temporal color video reconstruction from noisy CFA sequence. *IEEE Transactions on Circuits and Systems for Video Technology*, 20(6):838–847, 2010.

[399] L. Zhang, R. Lukac, X. Wu, and D. Zhang. PCA-based spatially adaptive denoising of CFA images for single-sensor digital cameras. *IEEE Transactions on Image Processing*, 18(4):797–812, 2009.

[400] L. Zhang and X. Wu. Color demosaicking via directional linear minimum mean square-error estimation. *IEEE Transactions on Image Processing*, 14(12):2167–2178, Dec. 2005.

[401] X.M. Zhang and B.A. Wandell. A spatial extension to CIELAB for digital color image reproduction. In *Proceedings of the SID Symposiums*, 1996.

[402] Y. Zhang, J. Xiao, and M. Shah. Motion layer based object removal in videos. *2005 Workshop on Applications of Computer Vision (WACV/MOTION)*, pages 516–521, 2005.

[403] Y. Zhang, D. Zhao, S. Ma, R. Wang, and W. Gao. A motion-aligned auto-regressive model for frame rate upconversion. *IEEE Transactions on Image Processing*, 19(5):1248–1258, 2010.

[404] W.Y. Zhao. Motion-based spatial-temporal image repairing. In *Proceedings of 2004 International Conference on Image Processing (ICIP)*, pages I: 291–294, 2004.

[405] W.Y. Zhao and H.S. Sawhney. Is super-resolution with optical flow feasible? In *Proceedings of the European Conference on Computer Vision (ECCV)*, pages 599–613. Springer, 2002.

[406] Z.R.H.J.H Zhiliang. Adaptive algorithm of auto white balance for digital camera. *Journal of Computer Aided Design & Computer Graphics*, 3:024, 2005.

[407] P. Zolliker and K. Simon. Retaining local image information in gamut mapping algorithms. *IEEE Transactions on Image Processing*, 16(3):664–672, 2007.

[408] R. Zone. The 3D zone. *Creative COW Magazine*, May/June 2009.

Index

T

U

V

W

Z